Robert C. Brears
**Water Resources Management**

# Also of Interest

*Hydrochemistry.*
*Basic Concepts and Exercises*
Eckhard Worch, 2023
ISBN 978-3-11-075876-4, e-ISBN 978-3-11-075878-8

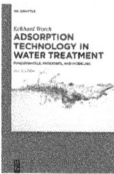

*Adsorption Technology in Water Treatment.*
*Fundamentals, Processes, and Modeling*
Eckhard Worch, 2021
ISBN 978-3-11-071542-2, e-ISBN 978-3-11-071550-7

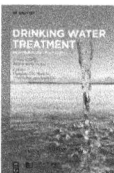

*Drinking Water Treatment.*
*New Membrane Technology*
Bingzhi Dong, Tian Li, Huaqiang Chu, Huan He, Shumin Zhu,
Junxia Liu (Eds.), 2022
ISBN 978-3-11-059559-8, e-ISBN 978-3-11-059684-7

*Water Resource Technology.*
*Management for Engineering Applications*
Vikas Dubey, Sri R.K. Mishra, Marta Michalska-Domańska,
Vaibhav Deshpande (Eds.), 2021
ISBN 978-3-11-072134-8, e-ISBN 978-3-11-072135-5

Robert C. Brears

# Water Resources Management

Innovative and Green Solutions

2nd Edition

**DE GRUYTER**

**Author**
Robert C. Brears
Our Future Water
79032 Avonhead
Christchurch 8446
New Zealand
rcb.chc@gmail.com

ISBN 978-3-11-102807-1
e-ISBN (PDF) 978-3-11-102810-1
e-ISBN (EPUB) 978-3-11-102928-3

**Library of Congress Control Number: 2023947863**

**Bibliographic information published by the Deutsche Nationalbibliothek**
The Deutsche Nationalbibliothek lists this publication in the Deutsche Nationalbibliografie;
detailed bibliographic data are available on the internet at http://dnb.dnb.de.

© 2024 Walter de Gruyter GmbH, Berlin/Boston
Cover image: Iwona Wozniak/iStock/Getty Images Plus
Typesetting: Integra Software Services Pvt. Ltd.
Printing and binding: CPI books GmbH, Leck

www.degruyter.com

# Acknowledgements

First and foremost, I wish to thank the team at De Gruyter who are visionaries and enable books like mine to come to fruition. Second, I would like to express my gratitude to my mum, who has a keen interest in environmental matters and has supported me in this journey of writing the book. Lastly, I want to extend a special acknowledgement to Kate, my love, who enriches my life in countless ways and has been an inexhaustible source of inspiration and motivation during the creation of this work.

https://doi.org/10.1515/9783111028101-202

# Contents

# Chapter 1
# Introduction

**Abstract:** As the century progresses, the water sector is facing increasing pressure from a wide range of climatic and non-climatic trends that challenge its ability to provide sustainable, reliable, resilient, and affordable water and water-related services that meet customers' expectations in the future. Traditionally, the water sector has been typically slow to evolve and incorporate new innovative solutions into existing systems in response to various challenges due to a number of barriers. Nonetheless, failure to implement innovations in water management will expose the water sector to a variety of risks. This book provides new research on innovative water management technologies that have been applied by leaders in the water sector to ensure the provision of sustainable, reliable, resilient, and affordable water and water-related services that meet customers' expectations in the future.

**Keywords:** Water Sector, Climate Change, Water Infrastructure, Innovation

As the century progresses, the water sector is facing increasing pressure from a wide range of climatic and non-climatic trends that challenge its ability to provide sustainable, reliable, resilient, and affordable water and water-related services that meet customers' expectations in the future.[1] Climate change will impact water resources in many ways, including increasing demand for scarce water while water supplies become limited, or variable, increasing the magnitude and frequency of storms and flooding, and increasing illnesses from poor water quality. Rapid population growth will increase the number of people living in water-scarce regions of the world while at the same time exposing a greater number of people to floods. Urbanisation is increasing demand for scarce water resources while at the same time impacting water quality in source watersheds. Rapid economic growth is placing significant demand on water for manufacturing and industrial processes. Furthermore, water scarcity is arising from growing water-energy and water-food nexus pressures. Ageing infrastructure is not only resulting in significant water losses from leakage but also deteriorating water quality. At the same time, the providing of water and wastewater-related services is contributing to rising greenhouse gas emissions. Meanwhile land-use change is impacting waterways, leading to biodiversity loss. Finally, customer expectations are rising with regards to the types of services they expect from water utilities, including greater emphasis on environmental sustainability.

The term 'water sector' is divided by the UN World Water Assessment Programme (WWAP) into three main functional categories: water resources management, which includes integrated water resources management and ecosystem restoration and remediation and is aimed at ensuring the protection, sustainable use, and regeneration

https://doi.org/10.1515/9783111028101-001

of water resources by protecting ecosystems, rivers, lakes, and wetlands, and building the necessary infrastructure, such as dams and aqueducts, to store water and regulate its flow; water infrastructure, which includes the construction, operation, and maintenance of water-related infrastructure, both human and natural, for the management of the resource as well as for the provision of water-related services, including the management of floods and droughts; and water services, which comprises the provision of services such as water supply, sanitation and hygiene, and wastewater management for domestic use as well as water-related services for economic use, for example, in energy, agriculture, and industrial sectors.[2]

Traditionally, the water sector has been typically slow to evolve and incorporate new innovative solutions into existing systems in response to various challenges due to a number of barriers.[3] Nonetheless, a failure to implement innovations in water management will expose the water sector to a variety of risks including environmental degradation, public health risks from poor quality water, damage to people and infrastructure from extreme weather events, and reductions in the level of service customers have come to expect.[4,5]

This book provides new research on innovative water management technologies that have been applied by leaders in the water sector to ensure the provision of sustainable, reliable, resilient, and affordable water and water-related services that meet customers' expectations in the future. In particular, the book provides readers with knowledge of how leaders in the water sector are implementing innovative technologies to conserve and recycle and reuse water, produce renewable energy and recover valuable nutrients from wastewater, protect and restore water quality at various scales, and improve the overall management of water resources. It also provides knowledge on the various innovative financial instruments and approaches available to meet water resources management challenges globally.

The synopsis of the book is as follows:

Chapter 2 discusses the various climatic and non-climatic challenges to the water sector before defining innovation. It will then discuss the various barriers to innovation before providing an overview of the strategies to overcome barriers to innovation.

Chapter 3 discusses how demand management utilises existing water supplies before plans are made to further increase supply before discussing water recycling and reuse innovations.

Chapter 4 discusses how wastewater treatment plants are water resource recovery facilities that produce clean water, reduce dependence on fossil fuels through the use and production of renewable energy, and recover nutrients.

Chapter 5 discusses the various green infrastructure solutions available to manage stormwater while utilising natural processes to improve water quality.

Chapter 6 provides an understanding of how river basin planning can protect and restore water quality before discussing how permit systems, best management practices, and source water protection can improve water quality.

Chapter 7 will first discuss the concept of smart digital water management followed by its components before discussing managing customers of the future.

Chapter 8 provides an overview of the various innovative financial instruments and approaches available to ensure the provision of sustainable, reliable, resilient, and affordable water and water-related services that meet customers' expectations in the future.

Chapter 9 provides an overview of best practices followed by the conclusion.

## Notes

1  UKWIR, "Research and Innovation Mapping Study for the UK Water Research and Innovation Framework.," (2011), https://www.theukwaterpartnership.org/research-and-innovation-mapping-study-for-the-uk-water-research-and-innovation-framework/.
2  United Nations World Water Assessment Programme, "Water and Jobs," (2016).
3  R.C. Brears, *Urban Water Security* (Chichester, UK; Hoboken, NJ: John Wiley & Sons, 2016).
4  Vanessa L. Speight, "Innovation in the Water Industry: Barriers and Opportunities for US and UK Utilities," *Wiley Interdisciplinary Reviews: Water* 2, no. 4 (2015).
5  R.C. Brears, *Climate Resilient Water Resources Management* (Cham, Switzerland: Palgrave Macmillan, 2018).

## References

Brears, R.C. *Climate Resilient Water Resources Management*. Cham, Switzerland: Palgrave Macmillan, 2018.
_____. *Urban Water Security*. Chichester, UK; Hoboken, NJ: John Wiley & Sons, 2016.
Speight, Vanessa L. "Innovation in the Water Industry: Barriers and Opportunities for US and UK Utilities." *Wiley Interdisciplinary Reviews: Water* 2, no. 4 (2015/07/01 2015): 301–13.
UKWIR. "Research and Innovation Mapping Study for the UK Water Research and Innovation Framework." (2011). https://www.theukwaterpartnership.org/research-and-innovation-mapping-study-for-the-uk-water-research-and-innovation-framework/.
United Nations World Water Assessment Programme. "Water and Jobs." (2016).

# Chapter 2
# Innovative water management

**Abstract:** The water sector is faced with multiple challenges in ensuring sustainable, reliable, resilient, and affordable water and water-related services that meet customers' expectations in the future. This chapter will first discuss the various challenges to the water sector before defining innovation and the multiple benefits of change to the water sector. The chapter will then survey the numerous barriers to innovation before providing an overview of the strategies the water sector can employ to overcome obstacles to change.

**Keywords:** Water Sector, Climate change, Water-Energy, Water-Food, Environmental Degradation, Innovation

## Introduction

The water sector is faced with multiple challenges in ensuring sustainable, reliable, resilient, and affordable water and water-related services that meet customers' expectations in the future. This chapter will first discuss the various challenges to the water sector before defining innovation and the multiple benefits of change to the water sector. The chapter will then survey the numerous barriers to innovation before providing an overview of the strategies the water sector can employ to overcome obstacles to change.

## 2.1 Multiple challenges to the water sector

The water sector is faced with multiple challenges in ensuring sustainable, reliable, resilient, and affordable water and water-related services that meet customers' expectations in the future, including the following.

### 2.1.1 Climate change

Climate change is projected to reduce renewable surface water and groundwater significantly. For each degree of global warming, around seven percent of the world's population is expected to be exposed to a decrease in renewable water resources of at least 20 percent. Climate change is likely to increase the frequency of meteorological droughts (less rainfall) and agricultural droughts (less soil moisture) in presently dry regions. The result is an increased frequency of short hydrological droughts (less sur-

https://doi.org/10.1515/9783111028101-002

face water and groundwater). Meanwhile, by the end of the century, around three times as many people will be exposed annually to the equivalent of a 20th-century 100-year river flood under a high emissions scenario (Representative Concentration Pathway 8.5 (RCP8.5)) than for a low emissions scenario (RCP2.6).[1] Climate change will impact water resources in many ways, including:

- *Increased demand for scarce water*: In many areas, climate change is likely to increase demand for water, while water supplies become limited, or variable. The result is water managers will need to simultaneously meet the needs of growing communities, sensitive ecosystems, farmers, energy producers, and manufacturers. In some areas, increased precipitation events and flooding will become more frequent, impacting the quality of water, and potentially leading to damage of infrastructure used to transport and deliver water
- *Increased storms and flooding*: More frequent and intense storms will overwhelm stormwater management systems, causing localised flooding and increasing run-off of contaminants, such as rubbish, nutrients, sediment, or bacteria, into local waterways. In urban centres with combined stormwater and wastewater drainage systems, more frequent and intense downpours can lead to combined sewer overflows into waterways, reducing water quality and making it difficult for cities to meet water quality standards
- *Increased illnesses from poor water quality*: Increased stormwater runoff into surface water bodies, indicated by increased turbidity from suspended solid particles eroded from the landscape, are associated with elevated levels of bacteria and other microorganisms. Small increases in the turbidity of drinking water have been linked to increased occurrence of acute gastrointestinal illnesses among children and the elderly
- *Eutrophication and algal blooms*: In lakes and reservoirs, there will be more intense eutrophication and algal blooms at higher temperatures, or shorter hydraulic retention times. During a harmful algal bloom, people can get exposed to toxins from fish they catch and eat, swimming in or drinking the water, and from the air they breathe. Depending on the type of algae, harmful algal blooms can cause serious health effects and even death, for example, eating seafood contaminated by toxins from algae called *Alexandrium* can lead to paralytic shellfish poisoning
- *Saltwater intrusion*: Sea level rise, together with increased groundwater pumping, can increase saltwater intrusion in groundwater aquifers. This can increase treatment costs for drinking water facilities or render groundwater wells unusable. Regarding surface water, sea-level rise may result in the 'salt front' (location of the freshwater-saltwater line) progressing further upstream. Saltwater intrusion of this nature can result in increased water treatment, relocation of water intakes, or the development of alternative sources of freshwater. Also, water infrastructure in coastal cities, including sewer systems and wastewater treatment facilities, are at risk from rising sea levels and the damaging impacts of storm surges[2,3,4,5,6,7,8,9]

### 2.1.2 Rapid population growth

The world's population is expected to increase from 7.7 billion currently to 9.7 billion in 2050, and potentially nearly 11 billion by 2100. The world's population will increase, but at varying rates globally. Nine countries will make up more than half the projected growth of the global population between now and 2050: India, Nigeria, Pakistan, the Democratic Republic of Congo, Ethiopia, the United Republic of Tanzania, Indonesia, Egypt, and the United States of America (in descending order of the expected increase). Regionally, the population of sub-Saharan Africa is estimated to more than double by 2050 (a 99 percent increase), while other regions will experience slower population growth: Oceania, excluding Australia/New Zealand (56 percent), Northern Africa and Western Asia (46 percent), Australia/New Zealand (28 percent), Central and Southern Asia (25 percent), Latin America and the Caribbean (18 percent), Eastern and South-eastern Asia (three percent), and Europe and Northern America (two percent).[10]

Already, around 3.6 billion people live in areas that experience water shortages for at least one month per year. By 2050, this could reach nearly six billion. Global demand for water has been increasing by around one percent per annum since the 1980s. Global demand for water is expected to continue rising at a similar rate until mid-century, accounting for an increase of 20 to 30 percent above the current level of water use, mainly due to increasing demand for water from the industrial and domestic sectors. At the same time, the number of people exposed to floods is expected to increase from 1.2 billion today to 1.6 billion in 2050.[11,12]

### 2.1.3 Urbanisation

Currently, 55 percent of the world's population reside in urban areas. By 2050, it is estimated that this will increase to 68 percent. Urbanisation is increasing in all regions of the world, but with a variation. Latin America and the Caribbean and North America are heavily urbanised, with over 80 percent of their population estimated to live in urban areas, rising to nearly 90 percent in 2050. Almost 75 percent of Europe's population is urban, and this is expected to rise to 85 percent by mid-century. Africa and Asia are urbanising more rapidly, with the percentage of Africa's and Asia's urban population likely to increase from approximately 40 percent and 50 percent respectively in 2018 to 59 percent and 66 percent in 2050.

While nearly an additional one billion persons will be added to today's urban population by 2030, more than half of the world's urban population will still be living in urban settlements with less than one million inhabitants, growing from 2.4 to 2.8 billion. The numbers of people living in medium-sized cities (1–5 million inhabitants) are expected to increase by 28 percent between now and 2030, growing from 926 million to 1.2 billion. Meanwhile, the number of megacities (cities with more than

10 million inhabitants) currently stands at 18. By 2030, the world is projected to have 43 megacities.[13]

Urbanisation is increasing demand for water resources with large cities estimated to obtain around 78 percent of their water from surface sources, some of which are far away. Cumulatively, large cities move 504 billion litres a day a distance of 27,000 ± 3,800 kilometres, and the upstream contributing area of urban water sources is 41 percent of the global land surface. Despite this infrastructure, one in four cities, containing $4.8 ± 0.7 trillion in economic activity, remain water stressed due to geographical and financial constraints.[14] Meanwhile, urbanisation is having an impact on water quality in urban source watersheds, with 90 percent of these watersheds having some level of degradation, with the average pollutant yield of municipal source watersheds increasing by 40 percent for sediment, 47 percent for phosphorus, and 119 percent for nitrogen. It is estimated that the degradation of watersheds has impacted treatment costs for 29 percent of cities globally, with operation and maintenance costs for affected cities increasing on average by around 53 percent and replacement capital costs rising by about 44 percent.[15]

### 2.1.4 Economic growth and resource use

Population growth, along with rising income levels, will drive a substantial increase in global demand for goods and services. Gross domestic product (GDP) is projected to quadruple between 2011 and 2060. By 2060, the global average per capita income is estimated to reach the current OECD level of around $40,000. Global materials use is projected to more than double from 79 gigatonnes (Gt) in 2011 to 167 Gt in 2060.[16] Between now and 2050, global demand for water is expected to increase for all major water use sectors: manufacturing (400 percent), thermal electricity generation (140 percent), and domestic use (130 percent). The most significant proportion of this growth in water demand is expected to occur in countries with developing or emerging economies, for example, in Africa water demand for industry will increase by 800 percent. In comparison, it will increase by 250 percent in Asia.[17,18] A failure to secure adequate and reliable supplies of water will result in the loss or disappearance of jobs. It is estimated that more than 1.4 billion jobs, or 42 percent of the world's total active workforce, are heavily dependent on water. Furthermore, it is estimated that 1.2 billion jobs, or 36 percent of the world's entire active workforce, are moderately water dependent. Overall, 78 percent of all jobs globally are dependent on water.[19]

### 2.1.5 Rising demand for energy

Energy is vital for a range of water processes, including water distribution, wastewater treatment, and desalination. Meanwhile, water is essential for all phases of energy production, including fossil fuels, biofuels, and power generation. In 2014, around four percent of global electricity consumption was used to extract, distribute, and treat water and wastewater, along with 50 million tonnes of oil equivalent of thermal energy, mostly diesel, used for irrigation pumps and gas in desalination plants. By 2040, the amount of energy used in the water sector is projected to more than double. The most substantial increase will come from desalination, followed by large-scale water transfer, and increasing demand for wastewater treatment and higher levels of treatment. For example, following a business-as-usual approach to centralised waste-water management, electricity consumption for urban municipal wastewater treatment could increase by over 600 terawatt-hours over the period to 2030.[20] Globally, the energy sector accounts for around 10 percent of total water withdrawals (the amount of water withdrawn from a source) and three percent of total water consumption (the volume of water withdrawn but not returned to the source). By 2040, it is projected that water withdrawals from the energy sector will increase by two percent to reach over 400 billion cubic metres (bcm). At the same time, the amount of water consumed is projected to increase by almost 60 percent to over 75 bcm, mainly due to a switch to advanced cooling technologies in the power sector that withdraw less water but consume more.[21]

### 2.1.6 Rising demand for food

Agriculture accounts for 70 percent of global freshwater withdrawals (surface and groundwater). Forty percent of irrigation uses groundwater sources, some of them non-renewable at the human time scale. It is estimated that food production will need to increase by 60 percent by 2050 to feed a population of over nine billion. Increased demand for food will result in irrigated food production growing by more than 50 percent by 2050. However, the amount of water withdrawn by agriculture can only increase by 10 percent, provided irrigation practices improve and yields rise. Furthermore, rising incomes and economic development are increasing demand for meat, fish, and dairy products, in addition to coarse grains and protein meals, impacting water resources with beef and dairy products being more water-intensive than cereals.[22,23]

The use of nitrogen and phosphorous, along with insecticides, herbicides, fungicides, and bactericides, in agricultural production is the leading cause of inland and coastal eutrophication. The result is algal blooms, loss of habitat and biodiversity, and long-term reduction or loss of fish catches. The runoff of farm and agro-processing chemicals into waterways and their seepage into aquifers poses risks to human health and the environment. Pollution reduces the availability of water for beneficial use and increases the cost of water treatment. Polluted water also has a high cost to human health with one-tenth of global disease attributed to water. Other pollution costs include clean-up and damage to fisheries, ecosystems, and recreation.[24] Furthermore, the current food production system relies on constant inputs of phosphorus to meet the growing demand for food. Phosphate rock, which is the dominant source of phosphorus for phosphate fertilisers, is a limited resource with current world reserves estimated to last between 30 and 300 years.[25]

### 2.1.7 Ageing infrastructure and deteriorating water quality

In many cities around the world, a large portion of the water infrastructure is approaching or has already reached the end of its useful life, with ageing infrastructure often resulting in high water loss from physical leakage.[26] In an earlier World Bank study, it was estimated that around 32 billion cubic metres of treated water physically leaks from urban water supply systems around the world each year.[27] Also, sewage, as well as contaminated groundwater surrounding pipes, can enter leaking pipes and travel throughout the water distribution network causing public health concerns, for example, outbreaks of gastrointestinal illness.[28,29] The required investment to rebuild these networks has to come on top of other water investment needs, including investments needed to comply with standards for drinking water quality. In the United States, it is estimated that repairing and expanding the country's drinking water infrastructure would top $1 trillion over 25 years and with increasing capital needs and potential funding shortfalls, many water utilities are increasing their rates they charge for water services in the immediate future.[30] However, there is evidence that customers' willingness to pay for any infrastructure upgrade is negatively affected by the cost of the proposed improvement.[31]

### 2.1.8 Greenhouse gas emissions

Water utilities are faced with climate change leading to increased water scarcity, lower water quality, and flooding challenges. At the same time, water utilities contribute to global emissions from energy consumption as well as nitrous oxide emissions and methane emissions from wastewater management. Water utilities collectively influence up to 12 percent of regional total primary energy consumption, with energy mainly used for water heating. Urban water utilities themselves typically account for one to two percent of aggregate global primary energy use and at times up to six percent of regional electricity use. The result is that the amount of regional greenhouse gas emissions contributed by urban water management is up to 17 percent.[32] It is estimated that 58 percent of emissions from urban water utilities comes from energy use while 40 percent is attributed to treatment processes and two percent from chemical use.[33]

### 2.1.9 Environmental degradation and biodiversity loss

Around the world, nature has been significantly altered by multiple human drivers with most indicators of ecosystems and biodiversity showing a rapid decline. Seventy-five percent of the land surface is changed substantially, 66 percent of the ocean is experiencing cumulative impacts, and over 85 percent of wetland areas have been lost. The average abundance of native species in most major terrestrial habitats have declined by at least 20 percent. Human actions threaten more species with global extinction now than ever before, with around 25 percent of species in the assessed animal and plant groups threatened. In total, approximately one million species are already facing extinction within decades unless action is taken to reduce the intensity of drivers of biodiversity loss. The main drivers of terrestrial and freshwater ecosystem degradation have been land-use change followed by the overexploitation of animals, plants, and other organisms mainly via harvesting, logging, hunting, and fishing.[34] The impacts of land-use change on water quantity and water quality include increased groundwater use from expanding human settlements; lower streamflow from land restoration activities; increased runoff resulting in increased sediment yield and higher nutrient loading of waterways from agricultural activities; increased total runoff and peak flow and a decrease in the baseflow; and evapotranspiration as a result of deforestation activities.[35,36,37]

### 2.1.10 High customer expectations

Water utilities are under increasing pressure to show customers the value for the rates paid and to enhance customer engagement and participation in various programmes.[38] The result is end-users of water services transitioning from being captive consumers of a uniform product delivered under fixed circumstances to end users that demand they be able to choose different products and services, for example, purchasing rainwater harvesting systems. Customers then turn from being consumers into co-constructors of new water infrastructure, helping to support water innovations while at the same time demanding these systems to be delivered and subsidised by the water utility or municipal agencies. Furthermore, water users are demanding that global water-using practices become more sustainable, which in turn provides support to water conservation initiatives developed by their local providers.[39]

## 2.2 Innovative water management

With challenges to the water sector increasing over the course of the century, there is an expectation that demand for innovative water management solutions will increase, in particular, solutions that enable the more efficient use of available water resources, enhance the quality of water for humans and nature, improve water resource planning to balance rising demand with limited, and often variable, supplies of water, and enhance resilience to extreme weather events.[40,41]

### 2.2.1 Innovation in the water sector

In the context of the water sector, innovation can be defined as *"the creation, development and implementation of a new product, technology, service, tariff design or process of production with the aim of improving efficiency, effectiveness or competitive advantage. It includes new ways of acquiring or deploying inputs, such as financial resources. The change may be incremental or fundamental."*[42] It should be noted that the definition includes the following:

- It deals with both products and processes
- It refers to the creation, development, implementation of a new product/process developed either in-house or by other companies and sectors
- All products and processes to be new or novel
- The aim must be to improve efficiency, effectiveness or increase competitive advantage[43]

#### 2.2.1.1 Degree of novelty

A vital aspect of the definition of innovation is that it must contain a degree of novelty, specifically one or more of the following:

- *New to the firm*: The innovation must be new to the firm. Other firms may have already implemented a product, process, marketing method or organisational method, but if it is new to the firm (or in the context of products and processes it is significantly improved), then it is an innovation for that firm
- *New to the market*: Innovations are new to the market when the firm is the first to introduce the innovation to the market, where a market is defined as the firm and its competitors, and it can include a geographic region or product line
- *New to the country*: An innovation is new to the country when the firm is the first to introduce the innovation for all domestic markets and industries
- *New to the world*: An innovation is new to the world when the firm is the first to introduce the innovation for all markets and industries internationally[44]

### 2.2.2 Innovative water management technologies

The term 'technology' is comprised of hardware, software, and orgware. Hardware includes physical infrastructure and technical equipment while software includes approaches, processes, and methodologies, for example, planning and decision support systems, models, knowledge transfer mechanisms, and capacity building. Orgware includes organisational and institutional arrangements as well as ownership models. There are four main categories of innovative water management technologies available to water managers in response to water sector challenges:

- *Supply enhancement*: Traditionally, water managers have met rising demand for water by increasing supply. However, with significant economic and environmental costs associated with supply-side management, water managers are increasingly focussing on innovations that create more drought-resilient water supplies, such as recycled water. Water managers are also focussing on decentralised systems such as rainwater and stormwater harvesting and on-site potable reuse systems. Furthermore, they are focusing on technologies that reduce energy use, such as extracting energy from wastewater, which in turn reduces water-energy nexus pressures

- *Demand management*: As water managers transition towards demand management, the focus will be increasingly on innovations that encourage or enable water conservation or water efficiency. Such innovations can decrease demand for new water supplies, increase water reliability, and decrease the costs and pollution associated with wastewater disposal. Innovations range from smart irrigation controllers to smart meters that encourage behaviour change

- *Green infrastructure*: Water managers can utilise natural processes to improve water quality and manage water quantity by restoring the hydrologic function of the landscape. Specifically, water managers can implement green infrastructure solutions at various scales to manage water quality and water quantity. Green infrastructure is defined as a strategically planned network of high-quality natural and semi-natural areas with other environmental features, which, in addition to managing water, is designed and managed to deliver a wide range of ecosystem services and protect biodiversity

- *Governance improvements*: Innovations can improve overall water governance, which is essential to securing access to reliable water supply and reducing demand. A wide range of innovations are available at various scales to reduce inefficiencies in the governance system, for example, smart grids can enable water utilities to quickly identify leaks in the distribution system while monitoring customer demand through smart meters can improve resource planning and management.[45,46,47,48]

### 2.2.3 Stakeholder contributions to innovative water management technologies

In addition to government entities at the local, state, and national level, many stakeholders within the water sector contribute to an innovative ecosystem including private sector companies and entrepreneurs, foundations, research centres, and trade associations. Through a variety of models, these actors can make progress towards developing and commercialising innovative water management technologies. The models include:

- *Public-private partnerships*: Partnerships with the private sector are an under-utilised tool in the water sector for meeting regulatory demands, making system improvements, and bringing new efficiencies and technologies to system operations. Public-private partnerships (PPP) can take many forms, including:
  - *Civil works and service contracts*: Utilities commonly source goods and services from private sector third parties, whether to purchase spare parts or to procure public works such as laying pipes. Utilities may also contract out a service, such as customer service. The utility will often purchase goods based on the provider's standard terms and conditions
  - *Management and operation and maintenance contracts*: These contracts govern the type of PPP agreement, which can range from technical assistance contracts to full-scale operation and maintenance agreements. Typically, the awarding authority engages the contractor to manage a range of activities for a relatively short period
  - *Build-Operate-Transfer (BOT) and Design-Build-Operate (DBO) projects*: A BOT project is typically used to develop a discrete asset rather than a whole network and is generally entirely new, or greenfield in nature, with the project company or operator obtaining its revenue through a fee charged to the utility/government. In a DBO project, the public sector owns and finances the construction of new assets while the private sector designs, builds, and operates the assets to meet certain agreed outputs
- *Public programmes*: At various levels of government, policymakers regularly re-shape institutions to meet interests, such as enhancing community wellbeing. Governmental actors may develop public policies to promote innovation to advance these interests, such as subsidising research and development, designing innovation prizes and challenges, creating publicly funded research laboratories, or providing marketplaces to foster strong collaborations and active networks
- *Regional collaboration*: Creating regional-wide water system collaborations can help provide the economies of scale and the technical, managerial, and financial capacity necessary for the development and adoption of water innovations. To reduce the fragmentation in the water sector and accelerate the adoption of innovative technologies, regional collaboration can be fostered by aligning technical experts, research institutions, and innovators regionally through non-profit organisations in the water sector that promote innovation through research, workshops, and collaboration as well as developing regional water clusters that connect water utilities with private partners and entrepreneurs.[49,50,51]

### 2.2.4 Barriers to innovative water management technologies

While water innovation provides many tangible benefits, including creating efficiencies, helping water systems meet regulatory requirements, and enabling better adaption to emerging pressures, there are many barriers to innovative water management technologies being developed, including:

- *Economic*: Water pricing is often not reflective of the costs of obtaining and transporting water. In many locations, water users are either charged a flat rate for water usage regardless of the volume used or a volumetric rate where the amount users pay is strictly based on the volume of water consumed, with neither pricing structure reflecting the rising costs of delivering higher amounts of water. The result is that revenue is often insufficient to cover the costs of infrastructure maintenance as well as investments in new water management technologies. Furthermore, water prices seldom reflect the costs of environmental damage
- *Financial*: The mainly public nature of the water sector is an initial barrier to available capital. Public entities commonly rely on bonds, issued at low-interest rates, to fund new projects. They are typically paid back using new revenue generated from the project or tapping into existing funds. However, rising operational and maintenance costs, as well as declining revenue from reduced demand from conservation efforts as well as leaks and inefficiencies in the water delivery system, threaten these funding sources. They can even affect bond ratings, further increasing the costs of new projects. This is particularly challenging for locations considering new technologies that might already present riskier rates of return than established technologies
- *Cultural and perception*: Change in regulations does not necessarily equate to innovation if other practices, norms, or cultural perceptions are not aligned. Substantial social and cultural barriers can inhibit current, proven technologies from being adopted, such as recycled water for direct potable reuse. Furthermore, the water sector is perceived to be less innovative than other sectors, resulting in less research and development investments being made as compared to other sectors
- *Institutional*: Institutions are broadly defined as the rules, norms, and practices that govern decision-making. This definition can include formal institutions, such as laws and regulations, as well as factors that shape water systems such as behaviour and cultural factors. Institutions may be a barrier to the uptake and utilisation of new technologies, for instance, there is often a reluctance to support unique/novel 'soft', sustainable technologies over traditional hard-engineering grey approaches
- *Infrastructure*: A lack of appropriate infrastructure can impede the development of innovations with current infrastructure being unable to support alternative practices. Often this is due to relying on conservative, highly visible infrastructure solutions rather than attempting to do new things. In addition, new technol-

ogy may not fit well into the existing system. In particular, new technologies may require complementary technologies that may not be available or are expensive or difficult to use

– *Lack of knowledge*: It is difficult for policymakers to keep up with science and technology relating to water management as the breadth and depth of water sector science and technology is significant. Furthermore, there are increasing research demands on limited available funds. Therefore, there is a lack of time and funds for policy and decision-makers to be familiar with the latest research findings and technology advancements

– *Risks adverse*: Water utilities are naturally risk-averse. As a water provider, the most considerable risk of applying new, innovative methods or technologies is creating an inadvertent disruption to the treatment or distribution of water. For example, with government agencies regulating the quality of water by setting baseline standards, violation of those standards will result in penalties from environmental agencies. Therefore, utilities are reluctant to try new approaches to treatment unless they are confident that the new technology will achieve the desired goals

– *Regulations*: Regulatory barriers are commonly cited as being one of the main barriers to water innovation, locking organisations into existing technologies. Regulatory regimes often develop around existing technologies and may clash with the characteristics of innovations. At times, manufacturers of existing technologies or other vested interests may use regulations as a market barrier. New technologies often face administrative costs stemming from the need for permits or other forms of regulatory approval that existing technologies do not face

– *Fragmentation*: In many countries, water management, infrastructure funding, and regulatory policies can differ within states and even within the same region or county. The result is a disjointed network that frequently prevents companies from establishing or spreading a new innovative technology. Also, with water utilities often operating autonomously, without an overarching and unifying body, any innovation must be independently tailored for each utility. For example, in the United States there are 3,200 electric utilities compared to 7,450 stormwater systems, over 16,000 publicly owned wastewater treatment systems, and more than 50,300 community water systems: therefore, achieving widespread adoption of any innovation in the water sector is a challenging task

– *Political*: Institutions often create barriers to the uptake of innovations due to lack of leadership or political will to initiate and sustain a transition towards new technology. For instance, water utility managers may lack significant support from superiors to initiate water conservation technology programmes with customers. Lack of political leadership or political will is often due to the lack of defined responsibility for decision-making or leadership lacking quality (skill set), integrity, transparency and accountability, coordination/interaction between government bodies, or capacity (financial and technical).[52,53,54,55,56,57,58,59,60,61,62,63,64,65,66,67]

### 2.2.5 Overcoming barriers

To overcome these barriers to innovation, the water sector's various actors can promote:

-   *Cultural change*: Innovation is about creating a culture and environment that allows changes to take hold and work in practice. It can also apply to the application of existing methods or technologies, in new ways or to new fields
-   *Collaboration*: Collaboration is essential for inspiring new ideas and applications, allowing for insights to develop, which in turn, spurs innovation. As well as collaborating with external stakeholders, water sector actors can collaborate within their organisations, with other organisations, and with partners outside of the sector
-   *Technology*: Technology, when paired with the right culture, processes, and people, is a powerful enabler of innovation. In addition to technology, such as smart meters and water-efficient appliances giving more control to water users over their consumption levels, technology can be applied to help water managers understand their systems and networks, helping them prevent interruptions to services and respond to and recover from service delivery challenges
-   *Innovative regulatory frameworks*: Regulatory frameworks can be designed to challenge various actors in the water sector to improve innovation for the benefit of customers, the environment, and broader society. Regulatory frameworks can encourage innovation by:
    -   Reconciling regulations that are inconsistent between government agencies and levels of government
    -   Coordinating regulations across sectors, for example, water and wastewater and water and energy, to ensure consistent treatment of new technologies and to reduce obstacles to the development and adoption of new technologies
    -   Shaping regulations to encourage utilities and various regulated water sector actors to meet performance standards, rather than force them to adopt fixed technology mandates
    -   Creating markets and competition in the water sector that encourages innovation through water trading, greater third-party involvement in large projects, and markets for ecosystem services
    -   Developing market-based instruments to recover the full cost of providing water and related services and encourage research and development in innovative projects in areas including water efficiency, resource recovery, and protection of ecosystems[68,69]

# Notes

1  B.E. Jiménez Cisneros, T. Oki, N.W. Arnell, G. Benito, J.G. Cogley, P. Döll, T. Jiang, and S.S. Mwakalila, Freshwater Resources, (Cambridge Cambridge University Press, 2014), https://www.ipcc.ch/site/assets/uploads/2018/02/WGIIAR5-Chap3_FINAL.pdf.
2  R.C. Brears, *Blue and Green Cities: The Role of Blue-Green Infrastructure in Managing Urban Water Resources* (Palgrave Macmillan UK, 2018).
3  Stephen J. Gaffield et al., "Public Health Effects of Inadequately Managed Stormwater Runoff," *American journal of public health* 93, no. 9 (2003).
4  National Institute of Environmental Health Sciences, "Harmful Algal Blooms," https://www.niehs.nih.gov/health/topics/agents/algal-blooms/index.cfm.
5  Bryony L. Townhill et al., "Harmful Algal Blooms and Climate Change: Exploring Future Distribution Changes," *ICES Journal of Marine Science* 75, no. 6 (2018).
6  US EPA, "Climate Adaptation and Saltwater Intrusion," https://www.epa.gov/arc-x/climate-adaptation-and-saltwater-intrusion.
7  A. Safi et al., "Synergy of Climate Change and Local Pressures on Saltwater Intrusion in Coastal Urban Areas: Effective Adaptation for Policy Planning," *Water International* 43, no. 2 (2018).
8  US EPA, "Climate Impacts on Water Resources," https://19january2017snapshot.epa.gov/climate-impacts/climate-impacts-water-resources_.html.
9  Thomas R. Allen et al., "Linking Water Infrastructure, Public Health, and Sea Level Rise: Integrated Assessment of Flood Resilience in Coastal Cities," *Public Works Management & Policy* 24, no. 1 (2018).
10  United Nations Department of Economic and Social Affairs, "World Population Prospects 2019: Highlights," (2019), https://population.un.org/wpp/Publications/Files/WPP2019_Highlights.pdf.
11  UN Water, "World Water Development Report 2019: Leaving No One Behind," (2019), https://knowledge.unccd.int/publications/world-water-development-report-2019-leaving-no-one-behind.
12  OECD, "Oecd Environmental Outlook to 2050: The Consequences of Inaction Highlights," (2012), https://www.oecd.org/g20/topics/energy-environment-green-growth/oecdenvironmentaloutlookto2050theconsequencesofinaction.htm.
13  United Nations Department of Economic and Social Affairs, "World Urbanization Prospects: The 2018 Revision," (2019), https://population.un.org/wup/Publications/Files/WUP2018-Report.pdf.
14  Robert I. McDonald et al., "Water on an Urban Planet: Urbanization and the Reach of Urban Water Infrastructure," *Global Environmental Change* 27 (2014).
15  Robert I. McDonald et al., "Estimating Watershed Degradation over the Last Century and Its Impact on Water-Treatment Costs for the World's Large Cities," *Proceedings of the National Academy of Sciences* 113, no. 32 (2016).
16  OECD, "Global Material Resources Outlook to 2060: Economic Drivers and Environmental Consequences," (2019), https://www.oecd.org/environment/waste/highlights-global-material-resources-outlook-to-2060.pdf.
17  UN-Water, "The United Nations World Water Development Report 2018: Nature-Based Solutions for Water," (2018), https://unesdoc.unesco.org/ark:/48223/pf0000261424.
18  Alberto Boretti and Lorenzo Rosa, "Reassessing the Projections of the World Water Development Report," *npj Clean Water* 2, no. 1 (2019).
19  UN-Water, "Water and Jobs," (2016), https://unesdoc.unesco.org/ark:/48223/pf0000244040/PDF/244040eng.pdf.multi.
20  IEA, "Water-Energy Nexus: World Energy Outlook Special Report," (2017), https://www.iea.org/reports/water-energy-nexus.
21  Ibid.
22  FAO, "Water for Sustainable Food and Agriculture: A Report Produced for the G20 Presidency of Germany," (2017), http://www.fao.org/3/a-i7959e.pdf.

**23** HLPE, "Water for Food Security and Nutrition. A Report by the High Level Panel of Experts on Food Security and Nutrition of the Committee on World Food Security," (2015), http://www.fao.org/3/a-av045e.pdf.

**24** FAO, "Water for Sustainable Food and Agriculture: A Report Produced for the G20 Presidency of Germany".

**25** Jessica G. Shepherd, Saran P. Sohi, and Kate V. Heal, "Optimising the Recovery and Re-Use of Phosphorus from Wastewater Effluent for Sustainable Fertiliser Development," *Water Research* 94 (2016).

**26** Ka Leung Lam, Steven J. Kenway, and Paul A. Lant, "Energy Use for Water Provision in Cities," *Journal of Cleaner Production* 143 (2017).

**27** World Bank, "The Challenge of Reducing Non-Revenue Water (Nrw) in Developing Countries: How the Private Sector Can Help: A Look at Performance-Based Service Contracting," (2006), https://sitere sources.worldbank.org/INTWSS/Resources/WSS8fin4.pdf.

**28** Sam Fox et al., "Experimental Quantification of Contaminant Ingress into a Buried Leaking Pipe During Transient Events," *Journal of Hydraulic Engineering* 142, no. 1 (2016).

**29** Melle Säve-Söderbergh et al., "Gastrointestinal Illness Linked to Incidents in Drinking Water Distribution Networks in Sweden," *Water Research* 122 (2017).

**30** AWWA, "Awwa's 2019 Water and Wastewater Rate Survey Reveals Increasing Utility Costs Boosting Rates," https://www.awwa.org/AWWA-Articles/awwas-2019-water-and-wastewater-rate-survey-reveals-increasing-utility-costs-boosting-rates.

**31** Eftila Tanellari et al., "On Consumers' Attitudes and Willingness to Pay for Improved Drinking Water Quality and Infrastructure," *Water Resources Research* 51, no. 1 (2015).

**32** WaCCLim, "The Roadmap to a Low-Carbon Urban Water Utility," (2018), http://wacclim.org/wp-con tent/uploads/2018/12/2018_WaCCliM_Roadmap_EN_SCREEN.pdf.

**33** Qian Zhang et al., "Hidden Greenhouse Gas Emissions for Water Utilities in China's Cities," *Journal of Cleaner Production* 162 (2017).

**34** IPBES, "Summary for Policymakers of the Global Assessment Report on Biodiversity and Ecosystem Services of the Intergovernmental Science-Policy Platform on Biodiversity and Ecosystem Serv ices," (2019), https://ipbes.net/sites/default/files/inline/files/ipbes_global_assessment_report_summary_for_policymakers.pdf.

**35** Pankaj Kumar et al., "Effect of Land Use Changes on Water Quality in an Ephemeral Coastal Plain: Khambhat City, Gujarat, India," *Water* 11, no. 4 (2019).

**36** Srilert Chotpantarat and Satika Boonkaewwan, "Impacts of Land-Use Changes on Watershed Discharge and Water Quality in a Large Intensive Agricultural Area in Thailand," *Hydrological Sciences Journal* 63, no. 9 (2018).

**37** Vinícius Augusto de Oliveira et al., "Land-Use Change Impacts on the Hydrology of the Upper Grande River Basin, Brazil," *CERNE* 24 (2018).

**38** NACWA, "Envisioning the Digital Utility of the Future," (2017), http://www.nacwa.org/docs/default-source/conferences-events/2017-summer/17ulc-digital-utility-r6.pdf?sfvrsn=2.

**39** D. L. T. Hegger et al., "Consumer-Inclusive Innovation Strategies for the Dutch Water Supply Sector: Opportunities for More Sustainable Products and Services," *NJAS – Wageningen Journal of Life Sciences* 58, no. 1 (2011).

**40** Uta Wehn and Carlos Montalvo, "Exploring the Dynamics of Water Innovation: Foundations for Water Innovation Studies," *Journal of Cleaner Production* 171 (2018).

**41** R.C. Brears, *Climate Resilient Water Resources Management* (Cham, Switzerland: Palgrave Macmillan, 2018).

**42** Vanessa L. Speight, "Innovation in the Water Industry: Barriers and Opportunities for Us and Uk Utilities," *Wiley Interdisciplinary Reviews: Water* 2, no. 4 (2015).

**43** Ibid.

**44** Ibid.

**45** UNEP-DTU UNEP-DHI Partnership, CTCN, "Climate Change Adaptation Technologies for Water: A Practitioner's Guide to Adaptation Technologies for Increased Water Sector Resilience," (2017), https://www.ctc-n.org/resources/climate-change-adaptation-technologies-water-practitioner-s-guide-adaptation-technologies.

**46** Barton Thompson Newsha Ajami, and David Victor, The Path to Water Innovation, (2014), https://www.hamiltonproject.org/papers/the_path_to_water_innovation.

**47** R.C. Brears, *Urban Water Security* (Chichester, UK; Hoboken, NJ: John Wiley & Sons, 2016).

**48** *Blue and Green Cities: The Role of Blue-Green Infrastructure in Managing Urban Water Resources.*

**49** World Bank, "Ppp Arrangements / Types of Public-Private Partnership Agreements," https://ppp.worldbank.org/public-private-partnership/agreements.

**50** Laura Diaz Anadon et al., "Making Technological Innovation Work for Sustainable Development," *Proceedings of the National Academy of Sciences* 113, no. 35 (2016).

**51** Bipartisan Policy Center, "Increasing Innovation in America's Water Systems," (2017), https://bipartisanpolicy.org/report/increasing-innovation-in-americas-water-systems/.

**52** Steven Greenland, "Sustainable Innovation Adoption Barriers: Water Sustainability, Food Production and Drip Irrigation in Australia," *Social Responsibility Journal* 15, no. 6 (2019).

**53** Michael Kiparsky et al., "Barriers to Innovation in Urban Wastewater Utilities: Attitudes of Managers in California," *Environmental Management* 57, no. 6 (2016).

**54** E. C. O'Donnell, J. E. Lamond, and C. R. Thorne, "Recognising Barriers to Implementation of Blue-Green Infrastructure: A Newcastle Case Study," *Urban Water Journal* 14, no. 9 (2017).

**55** Courtney Crosson, "Innovating the Urban Water System: Achieving a Net Zero Water Future Beyond Current Regulation," *Technology|Architecture + Design* 2, no. 1 (2018).

**56** Kimberly Duong and Jean-Daniel M. Saphores, "Obstacles to Wastewater Reuse: An Overview," *WIREs Water* 2, no. 3 (2015).

**57** Wehn and Montalvo, "Exploring the Dynamics of Water Innovation: Foundations for Water Innovation Studies."

**58** Bipartisan Policy Center, "Increasing Innovation in America's Water Systems".

**59** Mark F. Colosimo and Hyunook Kim, "Incorporating Innovative Water Management Science and Technology into Water Management Policy," *Energy, Ecology and Environment* 1, no. 1 (2016).

**60** Newsha Ajami, *The Path to Water Innovation.*

**61** Brears, *Urban Water Security.*

**62** Fox et al., "Experimental Quantification of Contaminant Ingress into a Buried Leaking Pipe During Transient Events."

**63** Säve-Söderbergh et al., "Gastrointestinal Illness Linked to Incidents in Drinking Water Distribution Networks in Sweden."

**64** Tanellari et al., "On Consumers' Attitudes and Willingness to Pay for Improved Drinking Water Quality and Infrastructure."

**65** Lam, Kenway, and Lant, "Energy Use for Water Provision in Cities."

**66** Michael Kiparsky et al., "The Innovation Deficit in Urban Water: The Need for an Integrated Perspective on Institutions, Organizations, and Technology," *Environmental Engineering Science* 30, no. 8 (2013).

**67** Brears, *Urban Water Security.*

**68** Ofwat, "Driving Innovation in Water," (2017), https://www.ofwat.gov.uk/publication/driving-innovation-water/.

**69** Newsha Ajami, *The Path to Water Innovation.*

# References

Allen, Thomas R., Thomas Crawford, Burrell Montz, Jessica Whitehead, Susan Lovelace, Armon D. Hanks, Ariel R. Christensen, and Gregory D. Kearney. "Linking Water Infrastructure, Public Health, and Sea Level Rise: Integrated Assessment of Flood Resilience in Coastal Cities." *Public Works Management & Policy* 24, no. 1 (2019/01/01 2018): 110–39.

Anadon, Laura Diaz, Gabriel Chan, Alicia G. Harley, Kira Matus, Suerie Moon, Sharmila L. Murthy, and William C. Clark. "Making Technological Innovation Work for Sustainable Development." *Proceedings of the National Academy of Sciences* 113, no. 35 (2016): 9682.

AWWA. "Awwa's 2019 Water and Wastewater Rate Survey Reveals Increasing Utility Costs Boosting Rates." https://www.awwa.org/AWWA-Articles/awwas-2019-water-and-wastewater-rate-survey-reveals-increasing-utility-costs-boosting-rates.

Bipartisan Policy Center. "Increasing Innovation in America's Water Systems." (2017). https://bipartisanpolicy.org/report/increasing-innovation-in-americas-water-systems/.

Boretti, Alberto, and Lorenzo Rosa. "Reassessing the Projections of the World Water Development Report." *npj Clean Water* 2, no. 1 (2019/07/31 2019): 15.

Brears, R.C. *Blue and Green Cities: The Role of Blue-Green Infrastructure in Managing Urban Water Resources.* Palgrave Macmillan UK, 2018.

——. *Climate Resilient Water Resources Management.* Cham, Switzerland: Palgrave Macmillan, 2018.

——. *Urban Water Security.* Chichester, UK; Hoboken, NJ: John Wiley & Sons, 2016.

Chotpantarat, Srilert, and Satika Boonkaewwan. "Impacts of Land-Use Changes on Watershed Discharge and Water Quality in a Large Intensive Agricultural Area in Thailand." *Hydrological Sciences Journal* 63, no. 9 (2018/07/04 2018): 1386–407.

Colosimo, Mark F., and Hyunook Kim. "Incorporating Innovative Water Management Science and Technology into Water Management Policy." *Energy, Ecology and Environment* 1, no. 1 (2016/02/01 2016): 45–53.

Crosson, Courtney. "Innovating the Urban Water System: Achieving a Net Zero Water Future Beyond Current Regulation." *Technology|Architecture + Design* 2, no. 1 (2018/01/02 2018): 68–81.

Duong, Kimberly, and Jean-Daniel M. Saphores. "Obstacles to Wastewater Reuse: An Overview." *WIREs Water* 2, no. 3 (2015/05/01 2015): 199–214.

FAO. "Water for Sustainable Food and Agriculture: A Report Produced for the G20 Presidency of Germany." (2017). http://www.fao.org/3/a-i7959e.pdf.

Fox, Sam, Will Shepherd, Richard Collins, and Joby Boxall. "Experimental Quantification of Contaminant Ingress into a Buried Leaking Pipe During Transient Events." *Journal of Hydraulic Engineering* 142, no. 1 (2016): 04015036.

Gaffield, Stephen J., Robert L. Goo, Lynn A. Richards, and Richard J. Jackson. "Public Health Effects of Inadequately Managed Stormwater Runoff." [In eng]. *American journal of public health* 93, no. 9 (2003): 1527–33.

Greenland, Steven. "Sustainable Innovation Adoption Barriers: Water Sustainability, Food Production and Drip Irrigation in Australia." *Social Responsibility Journal* 15, no. 6 (2019): 727–41.

Hegger, D. L. T., G. Spaargaren, B. J. M. van Vliet, and J. Frijns. "Consumer-Inclusive Innovation Strategies for the Dutch Water Supply Sector: Opportunities for More Sustainable Products and Services." *NJAS – Wageningen Journal of Life Sciences* 58, no. 1 (2011/06/01/ 2011): 49–56.

HLPE. "Water for Food Security and Nutrition. A Report by the High Level Panel of Experts on Food Security and Nutrition of the Committee on World Food Security." (2015). http://www.fao.org/3/a-av045e.pdf.

IEA. "Water-Energy Nexus: World Energy Outlook Special Report." (2017). https://www.iea.org/reports/water-energy-nexus.

IPBES. "Summary for Policymakers of the Global Assessment Report on Biodiversity and Ecosystem Services of the Intergovernmental Science-Policy Platform on Biodiversity and Ecosystem Services." (2019). https://ipbes.net/sites/default/files/inline/files/ipbes_global_assessment_report_summary_for_policymakers.pdf.

Jiménez Cisneros, B.E., T. Oki, N.W. Arnell, G. Benito, J.G. Cogley, P. Döll, T. Jiang, and S.S. Mwakalila,. *Freshwater Resources*. Cambridge Cambridge University Press, 2014. https://www.ipcc.ch/site/assets/uploads/2018/02/WGIIAR5-Chap3_FINAL.pdf.

Kiparsky, Michael, David L. Sedlak, Barton H. Thompson, and Bernhard Truffer. "The Innovation Deficit in Urban Water: The Need for an Integrated Perspective on Institutions, Organizations, and Technology." *Environmental Engineering Science* 30, no. 8 (10/24/received, 1/27/accepted 2013): 395–408.

Kiparsky, Michael, Barton H. Thompson, Christian Binz, David L. Sedlak, Lars Tummers, and Bernhard Truffer. "Barriers to Innovation in Urban Wastewater Utilities: Attitudes of Managers in California." *Environmental Management* 57, no. 6 (June 01 2016): 1204–16.

Kumar, Pankaj, Rajarshi Dasgupta, A. Brian Johnson, Chitresh Saraswat, Mrittika Basu, Mohamed Kefi, and K. Binaya Mishra. "Effect of Land Use Changes on Water Quality in an Ephemeral Coastal Plain: Khambhat City, Gujarat, India." *Water* 11, no. 4 (2019).

Lam, Ka Leung, Steven J. Kenway, and Paul A. Lant. "Energy Use for Water Provision in Cities." *Journal of Cleaner Production* 143 (2017/02/01/ 2017): 699–709.

McDonald, Robert I., Katherine F. Weber, Julie Padowski, Tim Boucher, and Daniel Shemie. "Estimating Watershed Degradation over the Last Century and Its Impact on Water-Treatment Costs for the World's Large Cities." *Proceedings of the National Academy of Sciences* 113, no. 32 (2016): 9117–22.

McDonald, Robert I., Katherine Weber, Julie Padowski, Martina Flörke, Christof Schneider, Pamela A. Green, Thomas Gleeson, *et al.* "Water on an Urban Planet: Urbanization and the Reach of Urban Water Infrastructure." *Global Environmental Change* 27 (2014): 96–105.

NACWA. "Envisioning the Digital Utility of the Future." (2017). http://www.nacwa.org/docs/default-source/conferences-events/2017-summer/17ulc-digital-utility-r6.pdf?sfvrsn=2.

National Institute of Environmental Health Sciences. "Harmful Algal Blooms." https://www.niehs.nih.gov/health/topics/agents/algal-blooms/index.cfm.

Newsha Ajami, Barton Thompson, and David Victor. *The Path to Water Innovation*. 2014. https://www.hamiltonproject.org/papers/the_path_to_water_innovation.

O'Donnell, E. C., J. E. Lamond, and C. R. Thorne. "Recognising Barriers to Implementation of Blue-Green Infrastructure: A Newcastle Case Study." *Urban Water Journal* 14, no. 9 (2017): 964–71.

OECD. "Global Material Resources Outlook to 2060: Economic Drivers and Environmental Consequences." (2019). https://www.oecd.org/environment/waste/highlights-global-material-resources-outlook-to-2060.pdf.

——. "OECD Environmental Outlook to 2050: The Consequences of Inaction Highlights." (2012). https://www.oecd.org/g20/topics/energy-environment-green-growth/oecdenvironmentaloutlookto2050theconsequencesofinaction.htm.

Ofwat. "Driving Innovation in Water." (2017). https://www.ofwat.gov.uk/publication/driving-innovation-water/.

Oliveira, Vinícius Augusto de, Carlos Rogério de Mello, Marcelo Ribeiro Viola, and Raghavan Srinivasan. "Land-Use Change Impacts on the Hydrology of the Upper Grande River Basin, Brazil." *CERNE* 24 (2018): 334–43.

Safi, A., G. Rachid, M. El-Fadel, J. Doummar, M. Abou Najm, and I. Alameddine. "Synergy of Climate Change and Local Pressures on Saltwater Intrusion in Coastal Urban Areas: Effective Adaptation for Policy Planning." *Water International* 43, no. 2 (2018/02/17 2018): 145–64.

Säve-Söderbergh, Melle, John Bylund, Annika Malm, Magnus Simonsson, and Jonas Toljander. "Gastrointestinal Illness Linked to Incidents in Drinking Water Distribution Networks in Sweden." *Water Research* 122 (2017/10/01/ 2017): 503–11.

Shepherd, Jessica G., Saran P. Sohi, and Kate V. Heal. "Optimising the Recovery and Re-Use of Phosphorus from Wastewater Effluent for Sustainable Fertiliser Development." *Water Research* 94 (2016/05/01/ 2016): 155–65.

Speight, Vanessa L. "Innovation in the Water Industry: Barriers and Opportunities for Us and Uk Utilities." *Wiley Interdisciplinary Reviews: Water* 2, no. 4 (2015/07/01 2015): 301–13.

Tanellari, Eftila, Darrell Bosch, Kevin Boyle, and Elton Mykerezi. "On Consumers' Attitudes and Willingness to Pay for Improved Drinking Water Quality and Infrastructure." *Water Resources Research* 51, no. 1 (2015): 47–57.

Townhill, Bryony L., Jonathan Tinker, Miranda Jones, Sophie Pitois, Veronique Creach, Stephen D. Simpson, Stephen Dye, Elizabeth Bear, and John K. Pinnegar. "Harmful Algal Blooms and Climate Change: Exploring Future Distribution Changes." *ICES Journal of Marine Science* 75, no. 6 (2018): 1882–93.

UN-Water. "The United Nations World Water Development Report 2018: Nature-Based Solutions for Water." (2018). https://unesdoc.unesco.org/ark:/48223/pf0000261424.

⎯⎯. "Water and Jobs." (2016). https://unesdoc.unesco.org/ark:/48223/pf0000244040/PDF/244040eng.pdf.multi.

UN Water. "World Water Development Report 2019: Leaving No One Behind." (2019). https://knowledge.unccd.int/publications/world-water-development-report-2019-leaving-no-one-behind.

UNEP-DHI Partnership, UNEP-DTU, CTCN,. "Climate Change Adaptation Technologies for Water: A Practitioner's Guide to Adaptation Technologies for Increased Water Sector Resilience." (2017). https://www.ctc-n.org/resources/climate-change-adaptation-technologies-water-practitioner-s-guide-adaptation-technologies.

United Nations Department of Economic and Social Affairs. "World Population Prospects 2019: Highlights." (2019). https://population.un.org/wpp/Publications/Files/WPP2019_Highlights.pdf.

⎯⎯. "World Urbanization Prospects: The 2018 Revision." (2019). https://population.un.org/wup/Publications/Files/WUP2018-Report.pdf.

US EPA. "Climate Adaptation and Saltwater Intrusion." https://www.epa.gov/arc-x/climate-adaptation-and-saltwater-intrusion.

⎯⎯. "Climate Impacts on Water Resources." https://19january2017snapshot.epa.gov/climate-impacts/climate-impacts-water-resources_.html

WaCCLim. "The Roadmap to a Low-Carbon Urban Water Utility." (2018). http://wacclim.org/wp-content/uploads/2018/12/2018_WaCCliM_Roadmap_EN_SCREEN.pdf.

Wehn, Uta, and Carlos Montalvo. "Exploring the Dynamics of Water Innovation: Foundations for Water Innovation Studies." *Journal of Cleaner Production* 171 (2018/01/10/ 2018): S1–S19.

World Bank. "The Challenge of Reducing Non-Revenue Water (Nrw) in Developing Countries: How the Private Sector Can Help: A Look at Performance-Based Service Contracting." (2006). https://siteresources.worldbank.org/INTWSS/Resources/WSS8fin4.pdf.

Wehn, Uta, and Carlos Montalvo. "Ppp Arrangements / Types of Public-Private Partnership Agreements." https://ppp.worldbank.org/public-private-partnership/agreements.

Zhang, Qian, Jun Nakatani, Tao Wang, Chunyan Chai, and Yuichi Moriguchi. "Hidden Greenhouse Gas Emissions for Water Utilities in China's Cities." *Journal of Cleaner Production* 162 (2017/09/20/ 2017): 665–77.

⎯⎯. "Ppp Arrangements / Types of Public-Private Partnership Agreements." https://ppp.worldbank.org/public-private-partnership/agreements.

Zhang, Qian, Jun Nakatani, Tao Wang, Chunyan Chai, and Yuichi Moriguchi. "Hidden Greenhouse Gas Emissions for Water Utilities in China's Cities." *Journal of Cleaner Production* 162 (2017/09/20/ 2017): 665–77.

# Chapter 3
# Conserving and recycling and reusing water

**Abstract:** Traditionally, water managers facing increased demand and variable levels of supply have relied on large-scale, supply-side infrastructural projects to meet increased demand for water. However, these projects are costly in both environmental and economic terms. Also, since most water resources are transboundary, supply-side projects can create political tensions. In contrast, demand management involves the better use of existing water supplies before plans are made to increase supply further. Meanwhile, water recycling and reuse water can increase supplies, which reduces the economic and environmental costs related to establishing new water supplies.

**Keywords:** Demand Management, Water Pricing, Water Metering, Water Restrictions, Water Recycling

## Introduction

Traditionally, water managers facing increased demand and variable levels of supply have relied on large-scale, supply-side infrastructural projects, such as dams and reservoirs, to meet increased demand for water (supply-side management). However, these projects are costly in both environmental and economic terms. Environmental costs include disruptions of waterways that support aquatic ecosystems. Economic costs stem from a reliance on more distant water supplies, often of inferior quality, which increases not only the costs of transportation but also the cost of treatment. Also, since the vast majority of water resources are transboundary, supply-side projects can create political tensions because they rely on water crossing both intra- and inter-state administrative and political boundaries. In contrast, demand management involves the better use of existing water supplies before plans are made to increase supply further. Meanwhile, water recycling and reuse can increase supplies, which reduces the economic and environmental costs related to establishing new water supplies. This chapter provides an overview of the numerous innovative demand management technologies available before discussing water recycling and reuse.

## 3.1 Demand management

Demand management promotes water conservation, during times of both normal conditions and uncertainty, through changes in practices, cultures, and people's attitudes towards water resources. In addition to the environmental benefits of preserving ecosystems and their habitats, demand management is cost-effective compared to supply-

https://doi.org/10.1515/9783111028101-003

side management because it allows the better allocation of scarce financial resources, which would otherwise be required to build expensive dams, water transfer schemes from one river basin to another, and desalination plants. Overall, demand management aims to:

– Reduce loss and misuse in various water sectors (intra-sector efficiency),
– Optimise water use by ensuring reasonable allocation between various users (cross-sectoral efficiency) while considering the supply needs of downstream ecosystems and other water users and uses
– Facilitate significant financial and infrastructural savings by minimising the need to meet increasing demand with new water supplies
– Reduce the stress on water resources by reducing or halting unsustainable exploitation of water resources

Demand management also involves developing alternative water supplies, which, in addition to conserving groundwater and surface resources, decreases diversion of freshwater from sensitive ecosystems, decreases discharge to sensitive water bodies, and saves energy.[1,2,3]

### 3.1.1 Demand management strategies

Demand management involves communicating ideas, norms, and innovations for water conservation across individuals and society, the purpose being to change people's culture, attitudes, and practices towards water resources and reduce consumption patterns. Water managers can use two types of demand management strategies to modify attitudes and behaviour towards water:

– Antecedent strategies attempt to influence the determinants of target behaviour before the performance of the behaviour
– Consequential strategies attempt to influence the determinants of target behaviour after the performance of the behaviour. This assumes that feedback, both positive and negative, of the consequences of that behaviour, will influence the likelihood of the behaviour happening or not happening in the future

## 3.2 Water pricing

A water tariff is a price assigned to water supplied by a public or private utility through a piped network to its customers. Water pricing is a long-used economic instrument to promote water conservation by creating disincentives for overuse. Economic theory suggests that water demand should behave like any other goods: as price increases, water use decreases. By serving as an incentive function, water pricing addresses water scarcity problems by promoting conservation as well as encourages investments in innovative, less water-intensive technologies. Also, water pricing can be used for cost recovery, with the water pricing scheme recovering direct costs (water supply and infrastructure costs) and indirect costs (environmental, social, and opportunity costs). There are two approaches to cost recovery: Supply cost recovery is the recovery of financial (internal) costs of water supply, including investments in infrastructure, operations and maintenance, and administrative costs. Full cost recovery is the recovery of financial as well as water use-related environmental, social, and opportunity costs. Overall, the meaning and level of cost recovery depend on what is considered to be part of the 'price' of providing and using water.

### 3.2.1 Common tariff structures

Frequently, a flat rate is charged for water usage regardless of the volume used, where typically the size of the charge is related to the customer's property value. In contrast, a volumetric rate is a charge based on the volume of water used at a constant rate. An increasing block tariff rate contains different prices for two or more pre-specified quantities (blocks) of water, with the price increasing with each successive block. A two-part tariff system involves a fixed and variable component. In the fixed component, water users pay one amount independent of consumption, which usually covers the administrative and infrastructural costs of supplying the water. Meanwhile, the variable amount is based on the quantity of water consumed and covers the costs of providing water as well as encouraging conservation.[4,5]

### 3.2.2 Irrigation tariff structures

Irrigation services can be charged for in many ways (Table 3.1). Each type of tariff provides different levels of incentive to irrigators to reduce consumption and different structures of income to the service provider.[6]

**Table 3.1:** Irrigation Water Tariffs.

| Type | Description |
| --- | --- |
| Area-based | A fixed rate per hectare of a farm, where the rate is not related to the area irrigated, the crop grown, or the volume of water received. It is usually part of a 'two-part' tariff designed to cover the fixed costs of the service. Different tariffs may be used for gravity and pumped supplies |
| | A fixed rate per hectare irrigated. The charge is not related to farm size, type of crop grown, or actual volume of water received (except that a larger irrigated area implies a greater volume of irrigation water) |

**Table 3.1** (continued)

| Type | Description |
|---|---|
| Crop-based | A variable-rate per irrigated hectare of the crop, i.e. different charges for different crops, where the charge is not related to the actual volume of water received, although the type of crop and area irrigated serve as proxies for the volume of water received |
| Volumetric | A fixed rate per unit water received, where the charge is related directly to, and proportional to, the volume of water received |
| | A variable-rate per unit of water received, where the service charge is related directly to the quantity of water received, but not proportionately (e.g. a certain amount of water per hectare may be provided at a low unit cost, a further defined quantity at a higher unit cost, and additional water above this further quantity at a very high unit cost). This method is referred to as a rising block tariff |
| Tradeable water rights | The entitlements of users in an irrigation project, or more widely, other users, are specified in accordance with the available water supply. Rights holders can buy or sell rights in accordance with specified rules designed primarily to protect the rights of third parties. Sales require authorisation by a licensing authority or may require court approval without reference to any specified authority |

**Case 3.1: Toronto's Industrial Water Rate Program**

Manufacturers in Toronto benefit from discounted water rates through the Industrial Water Rate Program, which aims to promote water conservation and support economic growth (also referred to as the Block 2 water rate).

To qualify for the program, manufacturers must meet the following conditions:
- Use over 5,000 cubic meters (m3) of water annually
- Fall under the industrial property tax classification
- Comply with Toronto's Sewers By-law
- Submit a comprehensive water conservation plan that meets the satisfaction of the General Manager of Toronto Water.

Manufacturers must follow these steps to apply for the programme:
- Step 1: Express your interest in the program by sending an email to Toronto Water
- Step 2: The City will confirm whether the manufacturer meets the eligibility criteria mentioned above. Even companies classified as "mixed" industrial and commercial may still be eligible.
- Step 3: If the company meets the eligibility requirements, they must prepare and submit a water conservation plan for approval. Customers who use up to 15,000 m3/year can also benefit from the CBB Program's free water audit, which can serve as a water conservation plan. Alternatively, customers can reach out to Partners in Project Green for help in developing their plan.
- Step 4: Enjoy the reduced water rates. If your conservation plan is approved:
  - The company will receive the Industrial Water Rate effective from the date of submitting the water conservation plan.
  - To maintain the discounted rate, the company must submit an annual water conservation progress report by July 1 of each year and continue to meet the other eligibility criteria.

Customers are charged the general water rate on the first 5,000 m3 of water use, while water usage exceeding that amount is eligible for a 30% rate reduction.

In 2023, the difference in water rates is as follows:
- General Water Rate (Block 1 Rate): $4.3863 per m3
- Industrial Water Rate (Block 2 Rate): $3.0703 per m3[7]

**Case 3.2: Santa Fe Irrigation District's Permanent Special Agricultural Rate**

Santa Fe Irrigation District (SFID) has recently launched the Permanent Special Agricultural Rate (PSAWR) on September 1, 2021. Customers who are currently part of the commercial agricultural program may opt to transition to the PSAWR program if they wish to do so. However, customers who opt not to transition will remain in the existing program. The PSAWR programme was initiated by the San Diego County Water Authority to acknowledge the economic significance of agriculture to the region. SFID will implement and manage this program, providing customers with reduced water costs in exchange for less reliability. In the event of a supply allocation under a shortage action plan, PSAWR customers may be requested to reduce their water use. Before initiating a cutback, SFID will contact customers, providing them with a timeline, parameters, and essential information regarding the requested reduction in use. The PSAWR rate will be $4.91 per HCF (748 gallons).[8]

## 3.3 Water metering

Automated Meter Reading (AMR) involves the automated transfer of recorded water consumption data via public or private radio to servers for the storage and subsequent processing of data by the utility and/or a third party. This usually involves the manipulation of existing manual meters, resulting in smart enabled meters. However, while AMR improves timeliness and accuracy of data, it does not significantly increase data density, for example, one read per month, although higher frequencies are possible. Advanced Metering Infrastructure (AMI) allows for two-way communication between the smart meter and the utility or other third party via the data logger as well as higher data density. AMI creates a data stream that enables real-time monitoring and analysis with high-resolution consumption data sent to the customer. This data can be used to raise awareness of water consumption and allow customers to develop their strategies to reduce water usage. From the water utility perspective, AMI meters provide multiple benefits including leak detection, energy reduction, demand forecasting, enhanced awareness campaigns, promotion of water-efficient appliances, and performance indicators. From the customer's perspective, smart meters provide information on when/where water is being used, comparisons of their water use against other customers, and quick leak detection. Water utilities can develop smart apps for customers to:
- Compare their water usage with neighbours in the same street or suburb
- Compare their consumption with standard profiles, such as consumers with the same socio-demographic factors

- Compare their water consumption with the most efficient users in the city
- Forecast their next water bill[9,10]

---

**Case 3.3: Severn Trent's Smart Meter Trial**

Severn Trent plans to install up to 1,800 smart water meters in West Bridgford, Nottingham to enhance its understanding of the water network and safeguard the local environment. The innovative trial will allow the company to obtain hourly data on water usage, enabling it to identify leaks faster and reduce water loss. The meters will be installed from March to April 2022, and the trial will run until early 2023. The smart meters will help the company to monitor water flow and detect leaks much faster, allowing them to rectify the situation before causing significant disruption. Severn Trent has to make assumptions regarding water usage on its network, but the smart meter trial will increase its comprehension of the network and whether increased water consumption is due to leaks or customer usage. Severn Trent reported that customers used an extra 60 litres of water per person on top of the usual 145 litres on the hottest day in summer 2021. The company aims to promote water conservation and tailor water efficiency initiatives to its customers by collecting data during the trial. The smart meters will also aid in identifying customers who could save money by switching to a metered bill. The trial will end in early 2023, and the installed meters will continue to send data to the company to inform Severn Trent's plans beyond 2025. Furthermore, Severn Trent launched a £20m Green Recovery Smart Metering program in Coventry and Warwickshire, aiming to install over 150,000 meters by 2025 to reduce leakage, save water for customers, and locate and rectify leaks quicker.[11]

---

## 3.4 Leak detection and water distribution network rehabilitation

In many distribution systems, up to half of the water supplied by the water treatment plant is lost to leakage. A significant part of the leaks occurring in a water distribution network does not reach the ground surface. These leaks can be detected by applying a range of active leakage control strategies, including
- Analysing changes in night inlet volume over time
- Setting flat-line alarm levels at crucial monitoring locations in a water distribution network, allowing near real-time identification of, usually, large bursts
- Using hydraulic sensor technology with utilities deploying many pressure and flow devices, with data coming from such devices. When used in combination with predictions of the water distribution network behaviour by hydraulic modelling, it has the potential to enable fast detection and location of pipe bursts

A water utility can improve the management and rehabilitation of its water distribution network with a well-planned maintenance programme based on sound knowledge of the distribution network. This knowledge is usually embodied in a distribution system database that includes the following data:
- An inventory of the characteristics of the system components, including information on their location, size, age, and the construction material(s) used in the network

- A record of regular inspections of the network including the condition of the mains and degree of corrosion
- An inventory of soil conditions and types, including the chemical characteristics of the soils
- A record of the quality of the product water in the system
- A record of any high- or low-pressure problems in the network
- Operating records, such as pump and valve operations, failures, leaks, and of maintenance and rehabilitation costs
- A file of customer complaints
- Metering data[12,13,14]

By monitoring these records, advanced warning of possible problems can be achieved. For example, numerous complaints could be a warning sign of an impending breakdown in the system. This system should also include a regular programme of preventative maintenance to minimise the possibility of system failures.[15]

Overall, leak detection and water distribution network rehabilitation programmes provide multiple benefits in addition to reducing water loss including:

- Increased revenue
- Reduced stress on the area's water resources
- Reduced energy consumption for abstraction, treatment, and distribution
- Improved water quality due to optimised water distribution as chlorine content in the distributed water will be better controlled and the risk of pollution related to bursts and periods with low pressure or vacuum will be reduced.[16]

**Case 3.4: Improving Leak Detection Efficiency in Sydney's Water Network**
The University of New South Wales (UNSW), with funding from Sydney Water, conducted a trial using new sonar technology to detect leaks in the city's 22,000 km of pipes. The existing acoustic approach to detect leaks, using copper wiring connected to microphones, was unsuitable for covering larger portions of the network. The new sonar array uses fibre optics, which is lighter, more agile, and less expensive. The sonar array consists of a 50-meter strip of garden hose with 16 microphones inside it, equally distributed along its length, to detect sounds and relay this information back to the software that locates a leak. The new technology, co-developed with Thales Underwater and UNSW-spinoff Zedelef, can be dragged through pipes by robots or embedded in new pipe sections for permanent leak detection. The trial has potential cost savings as it provides coverage to a larger part of Sydney Water's network. Moreover, it can catch leaks and repair pipes at an early stage. According to Sydney Water's 2017–18 Water Conservation Report, 129.5 megalitres of water are lost each day due to leaks, equivalent to almost 10% of the city's total water consumption. The trial was conducted before the introduction of water restrictions on June 1, 2019, due to dropping catchment dam levels.[17]

## 3.5 Water restrictions

There are two types of water restrictions as follows:
- *Temporary water conservation ordinances and regulations*: These restrict certain types of water use during specified times and/or restrict the level of water use to a specified amount. Examples of water-use regulations include:
  - Restrictions on non-essential water uses, for example, watering lawns, washing cars, filling swimming pools, and washing driveways
  - Restrictions on commercial use, for example, car washes, hotels, and other large consumers of water
  - Bans on using water of drinking quality for cooling purposes
- *Permanent water conservation ordinances and regulations*: These include amendments to building codes or ordinances requiring the installation of water meters and water-saving devices, for example:
  - Plumbing codes ensuring that all new homes and offices meet the maximum water-use standards for plumbing fixtures such as toilets, urinals, faucets, and showers
  - The requirement that low-flow toilets, showerheads, and faucets are installed in all newly constructed or renovated homes and offices[18]

### Case 3.5: Johannesburg's Water Restrictions
Johannesburg residents have been reminded by the city council that Level 1 water restrictions are still in place in the municipality. Johannesburg Water is committed to promoting water conservation and has been implementing these restrictions according to the Water Services By-law. The restrictions include no watering or irrigation of gardens between 06:00-18:00 from 1 September to 31 March and between 08:00-16:00 from 1 April to 31 August. Additionally, all consumers are prohibited from using a hose-pipe to clean paved areas and driveways with municipal water. As South Africa is a water-scarce country, the City of Johannesburg has urged all customers to use water sparingly, especially during the rain-starved winter months. In this regard, residents are advised to conserve water by implementing water-saving tips such as not leaving taps dripping, washing cars on the grass, using a watering can instead of a hosepipe, shortening showering time, using a glass of water to rinse when brushing teeth, and taking shallow baths. Furthermore, residents are encouraged to reuse water to water their gardens or pot plants. Finally, residents are advised to report any burst pipes, leaking water meters, or open hydrants. These measures will help the municipality to maintain a culture of water conservation and ensure sustainable use of the scarce resource.[19]

## 3.6 Water efficiency labelling

Water managers can promote water efficiency product labelling schemes that cover water-using devices such as taps, showers, and toilets. The labelling of household appliances according to their degree of water efficiency is essential in reducing household water consumption by eliminating unsustainable products from the market,

provided the labelling scheme is clear and comprehensible and identifies both the private and public benefits of conserving water. There are two main types of labelling schemes:
- *Endorsement labels*: The label indicates that a product has met a certain minimum standard
- *Rating labels*: The label indicates the level of efficiency by rating the product on a performance scale and/or by stating the product's actual water consumption or flow rate figures

Endorsement labels provide an easy tool for consumers to identify environmentally friendly or water-efficient products while rating labels provide a greater incentive for manufacturers to develop more efficient products and enable consumers to make more informed purchasing decisions. Both types can be either voluntary or mandatory and are often based on performance requirements and/or technical standards.[20,21]

**Case 3.6: Singapore's Mandatory Water Efficiency Labelling Scheme**
Singapore's Mandatory Water Efficiency Labelling (Mandatory WELS) scheme, which began in 2009, is a grading system that assigns 0/1/2/3 tick ratings to products based on their water efficiency level. This system was introduced as a follow-up to the Voluntary Water Efficiency Labelling Scheme (Voluntary WELS), which was implemented in 2006. Currently, the Mandatory WELS covers taps and mixers, dual-flush low capacity flushing cisterns, urinal flush valves, and waterless urinals (with 2/3 tick ratings) as well as clothes washing machines for household use (with 2/3/4-tick ratings). Suppliers and retailers are required to obtain the relevant water efficiency labels for their products before advertising and displaying them for sale in Singapore, and all products must publicly display their water efficiency label at all times. This allows consumers to make informed choices when making purchases. Effective 1 January 2022, the Mandatory WELS system will be extended to include WC flush valves with a flush volume of not more than 4.0 litres per flush, and only WC flush valves with 2-tick ratings or higher will be allowed for supply. Additionally, only commercial washer extractors, commercial dishwashers, and high pressure washers that are registered under the Mandatory WELS and meet the mandatory water efficiency requirements will be permitted for supply in Singapore starting 1 January 2022. Furthermore, as of 1 October 2018, the Mandatory WELS has expanded its product range to include dishwashers for household use. This expansion aims to promote water conservation and increase awareness of water efficiency among consumers.[22]

## 3.7 Education and awareness

Education and awareness tools aim to change behaviour through public awareness campaigns around the need to conserve scarce water resources. Water utilities can promote water conservation in schools to increase young people's knowledge of the water cycle and encourage the wise use of scarce water resources. Water utilities can use a variety of strategies, including school presentations and distribution of water conservation information and materials that can be used in the school curriculum. Meanwhile, water utilities can use public education to persuade individuals and communities to conserve water resources. Water utilities can influence an individual's attitudes and behaviours towards water resources by increasing their knowledge and awareness of environmental problems associated with water scarcity. There are multiple tools and formats that water utilities can use to increase environmental awareness and water conservation, including:

- Public information such as television commercials, newspaper articles and advertisements
- Internet and social media campaigns
- Public events such as conservation workshops, public exhibitions
- Information included in water utility bills, such as leaflets on water conservation tips

Education and awareness campaigns can also involve the distribution of water conservation kits and providing of rebates to encourage physical water savings. Overall, a variety of best practices can be followed to ensure education and awareness campaigns are successful:

- Campaigns are most effective when they use a well-targeted range of media
- The use of existing networks can lower the cost of campaigns and increase their impact
- The provision of information needs careful management to ensure it is relevant and credible
- The impact of a water campaign can be magnified if it is followed by tangible action[23,24,25,26,27,28]

**Case 3.7: Scottish Water's Education Resources**

Scottish Water has developed a range of educational resources for schools, including:

- *"All About Water"*: A section where students can learn about the water cycle, our bodies and water, the environment and saving water, and much more.
- *Modules*: An education programme supporting learning and teaching within Scotland's school curriculum. Modules are developed for Early, First, Second, Third, and Fourth levels, and can be downloaded and supported with additional learning resources. Teachers/community group leaders can select which topics and activities they are interested in and tailor the topic or activity. Some activities are more classroom or indoor-based, while others can be done outside or at home.
- *Games*: Fun water and waste water related games to support learning, including:
  - *Clean it Up*: A waste water treatment game where players must remove items like wipes, cotton buds, nappies, and used fat, oil, and grease from the waste water before it's released into the environment. The game is aimed at teaching students about proper waste disposal and its effects on the environment.
  - *Pipeline Challenge*: A five-level game where players must build a water supply and waste water network by laying pipes between reservoirs, towns, and treatment works while avoiding obstacles along the way. The game is designed to help students understand the challenges of managing a water network and the importance of proper maintenance.

  - *Pumping Station*: A game where players experience the effort required to supply fresh, clear water by pumping it to supply virtual residents. The game encourages players to consider how to be more efficient with water use, which can help save both water and energy.
- *Digital Learning Hubs*: two new digital learning hubs aimed at both Primary and Secondary school pupils, designed to teach about the Water Cycle, how it is affected by climate change, and how Scottish Water works with nature's cycle. The Primary School Education Hub is aimed at adults who can register and share content with classes or children, while the Secondary School education hub is aimed at adults or any users over the age of 13.
- *Get to know H20*: a brand new Water Cycle Education Kit for parents, carers, and teachers across Scotland, created in response to the temporary closure of all Scottish schools due to the Covid-19 pandemic. The online water cycle learning pack aims to teach pupils about the water cycle and how they can help protect the environment. The pack also encourages pupils to think about the skills they have which could lead them to a future career in engineering, communication and more.[29]

# 3.8  Demonstration projects

Demonstration projects illustrate the feasibility and commercial viability of water conservation and water efficiency initiatives. They can also showcase the various economic, environmental, and social benefits to the community, including water utility customers, private building owners, and developers. Furthermore, demonstration projects can pilot new policies for municipal or regional governments and build local and institutional capacity and confidence.[30]

**Case 3.8: Sydney Water's Purified Recycled Water Demonstration Plant**
The Greater Sydney Water Strategy, recently published by the government of New South Wales, highlights the urgent need to invest in unconventional water sources to ensure water supply resilience in the face of climate change and population growth. The strategy proposes increasing water conservation and efficiency, as well as planning for new rainfall-independent supply sources, such as desalination and large-scale wastewater recycling. However, with the recycling of water for drinking water being a contentious issue, the state's minister for water emphasised the less contentious option of using recycled water for greening and cooling purposes. In response, Sydney Water is building a Purified Recycled Water Demonstration Plant and visitor centre to showcase the available technology to produce high-quality recycled water for greening and cooling projects, such as waterways and environmental flows. The plant will treat filtered wastewater from the Quakers Hill Water Recycling Plant, and the visitor centre will educate the public on purified recycled water. The plant will not be added to Sydney's drinking supply and is expected to be completed by early 2023 within the existing Sydney Water site to minimize community impact.[31]

## 3.9 Water recycling and reuse

Water recycling is the use of harvested water for the same or a different function, after treatment, where treatment can be tailored to meet the water quality requirements of planned use. In contrast, water reuse is the direct use of harvested water for the same or a different function, without treatment. Water recycling and water reuse systems can provide a reliable, climate-resilient, and economically sound source of water for non-potable and potable uses.[32,33]

### 3.9.1 Non-potable use

Water recycling and water reuse systems can be developed for a variety of non-potable projects. The water is usually of a lower quality than potable systems, and the level of treatment varies depending on the end-use.[34]

#### 3.9.1.1 Industrial
Industrial water can be reused or recycled within a business itself or between several businesses. A business can directly reuse wastewater that is clean enough for the purpose for which it is being reused. Process water is produced by industrial processes such as cooling and heating and usually contains few contaminants, often making it suitable for reuse. Cooling towers are one of the most common water technologies in use by industry and the water is frequently reused for washing processes. Industrial process water and cooling tower water can be treated to meet fit-for-purpose specifications for a particular next use. Meanwhile, water recycling or reuse systems can be implemented for use between businesses, with the exchange of waste product for the mutual benefit of two or more businesses known as 'industrial symbiosis'.

#### 3.9.1.2 Agricultural
The increased availability of treated (secondary-treated wastewater) and recycled water (tertiary-treated), along with increased competition for water supplies, provides an opportunity to develop this resource for agricultural production particularly during times of drought when regular water supplies are limited or non-existent. The use of treated and recycled water for irrigated crop production is controlled by regulations that govern the treated water quality, with lesser standards required for forage crops compared with those for food crops.[35]

#### 3.9.1.3 Urban
In urban areas, a variety of onsite non-potable reuse and recycling systems are utilised to meet non-potable needs:

- *Greywater*: Greywater is reusable wastewater from residential, commercial, and industrial bathroom sinks, bathtub shower drains, and clothes washing equipment drains. Greywater is reused onsite, usually for toilet flushing and irrigation. Greywater systems vary significantly in their complexity and size, ranging from small systems with simple treatment processes to large systems with complex treatment processes. Nevertheless, most have standard features including a tank for storing the treated water, a pump, a distribution system for transporting the treated water to where it is needed, and some sort of (basic) treatment, such as filtering, settlement of solids, chemical or UV disinfection etc.
- *Blackwater*: Blackwater, or sewage, is the wastewater from toilets. In blackwater recycling systems, all the blackwater is routed to an initial tank via gravity, from which it settles, and a primary colony of bacteria eats at the waste. The blackwater then goes through an aeration stage and a sludge settling stage, before it is chlorinated and used as irrigation water (watering lawns or non-food gardens)
- *Rainwater harvesting*: Rainwater harvesting systems collect and store rainfall for later use. When designed appropriately, they slow down and reduce runoff and provide a source of water. There are two main types of rainwater harvesting systems:
  - *Passive harvesting systems*: They are typically small volume (50–100 gallon) systems designed to capture rooftop runoff. Rain barrels are usually used in residential applications where the flow from rain gutter downspouts is easily captured for outdoor use, for example, garden and landscape irrigation or car washing
  - *Active harvesting systems*: They are larger volume (typically 1,000–100,000-gallon) systems, for example, cisterns, which capture runoff from roofs or other suitable surfaces. Rainwater collected in active systems is typically used for irrigation or indoor non-potable water replacement, for example, toilet flushing, clothes washing, evaporative cooling, etc.[36]

**Case 3.9: The HAMBURG WATER Cycle® in the Jenfelder Au Quarter**
The Jenfelder Au quarter is a new residential area in the Wandsbek district of Hamburg, built on the former Lettow-Vorbeck military barracks. The HAMBURG WATER Cycle® (HWC) will be implemented in over 800 newly constructed housing units and entails the separation of the material flows of wastewater. In conventional systems, all domestic wastewater streams are combined and discharged together into the sewer system. In contrast, the HWC decouples the wastewater flows. Blackwater, greywater, and stormwater are separately collected and then separately treated:
- *Blackwater*: Blackwater is less dilute and therefore facilitates material and energy recovery as well as reduces the energy required to treat it. To further concentrate the blackwater, water-saving toilets are used. The vacuum toilet consumes only about one litre of water per flush, saving around five to nine litres of water per flush when compared with conventional toilets. The concentrated blackwater is then combined with other biomass sources such as organic waste in anaerobic digesters. Biogas is then formed which is converted into electricity and heat from a combined heat and power process

- *Greywater.* In the HWC, greywater is separated from blackwater and is transported to a specially designed facility before it is introduced into the local waters. They greywater can also be used as process water for household activities such as watering the garden or toilet flushing
- *Stormwater.* The HWC aims to manage stormwater on-site and as natural as possible. The stormwater can be used for watering lawns or managed using decentralised methods, such as retention ponds, where water is able to either evaporate or join nearby waters, improving the local climate or recharging groundwater[37]

**Case 3.10: Hunter Water's recycled water service**
Hunter Water in Australia is using recycled water as one approach to ensuring the reliable delivery of water to the Lower Hunter Region while ensuring confidence in public health and environmental protection. Over the period 2017–18, the water utility recycled 6,454 million litres of wastewater. Hunter Water's recycled water scheme can be divided into four categories:
- *Open space*: Recycled water is used for open space irrigation across the Hunter, including several golf courses and an educational college. These schemes use around 500 million litres of recycled water per year for landscape irrigation
- *Residential*: Hunter Water operates two dual reticulation schemes that provide recycled water for garden and toilet flushing in several housing estates. The water utility is also investigating dual reticulation schemes for other new residential developments
- *Industrial use*: Industrial reuse customers such as Eraring Power Station and the Oceanic Coal Washery use around 1,600 million litres of recycled water per year
- *Agriculture use*: This includes local farmers in the Clarence Town Irrigation Scheme which is an integral part of the Clarence Town Wastewater Treatment Works, which is designed to recycle all the product effluent in dry years. The final recycled water product is then stored in a 34 million litre dam before being pumped to the reuse area. The reuse area consists of 18 hectares of pasture irrigated by a system of irrigators. Commercial fodder crops are cultivated on the irrigated area[38]

## 3.9.2 Potable reuse

Potable water reuse involves the use of a community's wastewater as a source of drinking water. Two forms of planned potable reuse exist, which are indirect potable reuse (IPR) and direct potable reuse (DPR).

### 3.9.2.1 Indirect potable reuse
IPR can be defined as the reclamation and treatment of water from wastewater (often sewage effluent) and the eventual returning of it into the current/natural water cycle well upstream of the drinking water treatment plant. Planned IPR means there is an intent to reuse the water for potable use. The point of return could be either into a major water supply reservoir; a stream feeding a reservoir; or into a water supply aquifer (managed aquifer recharge). The natural processes of filtration and dilution of the water with natural flows aim to reduce real or perceived risks associated with eventual potable reuse. IPR (unplanned) is defined as treated wastewater entering the natural water (creeks, rivers, lakes,

aquifers), which is eventually extracted from the natural system for drinking water: usually with no awareness that the natural system contains treated wastewater.[39]

**Case 3.11: Orange County's Groundwater Replenishment System**
Orange County's Groundwater Replenishment System (GWRS), a joint project between the Orange County Water District and the Orange County Sanitation District (OCSD), provides enough new water for nearly 850,000 residents and has become an essential local water supply. The GWRS is the world's most extensive advanced water purification system for IPR. OCSD treats the wastewater and produces water clean enough to undergo purification at the GWRS. The water is purified at the GWRS using a three-step advanced process of microfiltration, reverse osmosis, and ultraviolet light with hydrogen peroxide. The purified water is then injected into a seawater barrier and pumped to recharge basins where it percolates into the Orange County Groundwater Basin and supplements Orange County's drinking water supplies. Operational since 2008, the GWRS originally produced 70 million gallons a day (MGD) of purified water. In 2015, the project was expanded to produce 100 MGD. The GWRS' total capacity is projected to reach 130 MGD after the infrastructure is built to increase wastewater flows from OCSD to the GWRS, with the final expansion expected to be completed in 2023.[40]

## 3.9.2.2  Direct potable reuse

DPR can be defined as either the injection of recycled water directly into the potable water supply distribution system downstream of the water treatment plant or into the raw water supply immediately upstream of the water treatment plant (injection could be either directly into the water pipeline or a service reservoir). This means water used by consumers could contain either undiluted or slightly diluted recycled water. The key distinction with IPR is that there is no temporal or spatial separation between the recycled water introduction and its distribution to consumers.[41,42,43]

**Case 3.12: Windhoek's direct potable reuse system**
Windhoek, Namibia, is the first city in the world to produce drinking water directly from the municipal wastewater. For over 50 years, the city has been producing DPR with the first plant commissioned in 1968 and the second in 2001. Each day, 21,000 cubic metres of potable water is produced for direct reuse. There are five main aspects of the project:
1.  *Multi-barrier approach*: A multi-barrier approach is taken to treat the water with the fundamental processes being powdered activated carbon dosing, pre-oxidation and pre-ozonation, flash mixing, enhanced coagulation and flocculation, dissolved air flotation, dual media rapid gravity sand filtration, ozonation, biological activated carbon filtration, granular activated carbon filtration, ultrafiltration, and disinfection and stabilisation
2.  *Guaranteed water quality values*: The water produced must adhere to 'guaranteed values' based on World Health Organization Guidelines, Rand Water Potable Water Quality Criteria, and the Namibian Guidelines for Group A water. Water samples are taken every four hours at various points throughout the plant and analysed for basic quality control purposes
3.  *Blending of recycled and freshwater*: Blending the recycled water with treated surface water and/or groundwater provides an additional level of safety. The maximum portion of recycled water fed into the distribution system is 50 percent in times of water scarcity and low water demand

4.  *Operation and maintenance agreement*: The plant is operated and maintained under a twenty-year operation and maintenance contract between the City of Windhoek and a consortium of three international water treatment contractors
5.  *Public awareness campaign*: Persistent, well designed, and targeted marketing has meant the people of Windhoek generally take pride that they are the only city in the world where DPR is practised[44,45]

## Notes

1  R.C. Brears, "Urban Water Security in Asia-Pacific: Promoting Demand Management Strategies," (2014), https://refubium.fu-berlin.de/bitstream/handle/fub188/18349/pp414-urban-water-security-asiapa cific.pdf?sequence=1&isAllowed=y.

2  *Urban Water Security* (Chichester, UK; Hoboken, NJ: John Wiley & Sons, 2016).

3  *Developing the Circular Water Economy* (Cham, Switzerland: Palgrave Macmillan, 2020).

4  "Urban Water Security in Asia-Pacific: Promoting Demand Management Strategies".

5  European Commission, "The Role of Water Pricing and Water Allocation in Agriculture in Delivering Sustainable Water Use in Europe – Final Report" (2012), https://ec.europa.eu/environment/water/quantity/pdf/agriculture_report.pdf.

6  FAO, "Water Charging in Irrigated Agriculture," http://www.fao.org/3/y5690e/y5690e04.htm

7  Toronto Water, "Industrial Water Rate Program," https://www.toronto.ca/services-payments/water-environment/how-to-use-less-water/water-efficiency-for-business/industrial-water-rate-program/#:~:text=The%20Industrial%20Water%20Rate%20Program,m3)%20of%20water%20annually.

8  SFID, "Permanent Special Agricultural Water Rate Program (Psawr)," https://www.sfidwater.org/288/Permanent-Special-Agricultural-Water-Rat#:~:text=The%20uniform%20rate%20for%20all%20agricultural%20water%20use,the%20regular%20rate%20%28s%29.%20Frequently%20Asked%20Questions%20%28link%29.

9  Thomas Boyle et al., "Intelligent Metering for Urban Water: A Review," *Water* 5, no. 3 (2013).

10  C. D. Beal and J. Flynn, "Toward the Digital Water Age: Survey and Case Studies of Australian Water Utility Smart-Metering Programs," *Utilities Policy* 32 (2015).

11  Severn Trent, "Severn Trent to Begin New Smart Meter Trial in West Bridgford to Combat Leakage," https://www.stwater.co.uk/news/news-releases/severn-trent-to-begin-new-smart-meter-trial-in-west-bridgford-to/.

12  S.N. Ghosh, *Environmental Hydrology and Hydraulics: Eco-Technological Practices for Sustainable Development* (CRC Press, 2016).

13  Mahmud Güngör, Ufuk Yarar, and Mahmut Firat, "Reduction of Water Losses by Rehabilitation of Water Distribution Network," *Environmental Monitoring and Assessment* 189, no. 10 (2017).

14  Luigi Berardi et al., "Active Leakage Control with Wdnetxl," *Procedia Engineering* 154 (2016).

15  UNEP, "Source Book of Alternative Technologies for Freshwater Augmentation in Latin America and the Caribbean," https://www.oas.org/dsd/publications/Unit/oea59e/ch20.htm#2.1%20desalination%20by%20reverse%20osmosis.

16  State of Green, "Reducing Urban Water Loss," (2016), https://stateofgreen.com/en/publications/reducing-urban-water-loss/.

17  iTnews, "Sydney Water Taps Unsw Fibre Optic Sonar to Find Leaks," https://www.itnews.com.au/news/sydney-water-taps-unsw-fibre-optic-sonar-to-find-leaks-525862.

18  Brears, "Urban Water Security in Asia-Pacific: Promoting Demand Management Strategies".

19  City of Johannesburg, "Level 1 Water Restrictions Remain in Place for Joburg," https://www.joburg.org.za/media_/Newsroom/Pages/2019%20Newsroom%20Articles/April%202019/Level-1-water-re

strictions-remain-in-place-in-Joburg.aspx#:~:text=Johannesburg%20Water%2C%20in%20an%20effort%20to%20maintain%20a,·%20Do%20not%20leave%20taps%20dripping.%20More%20items.

20   Brears, "Urban Water Security in Asia-Pacific: Promoting Demand Management Strategies".

21   D. A. Kelly, "Labelling and Water Conservation: A European Perspective on a Global Challenge," *Building Services Engineering Research & Technology* 36, no. 6 (2015).

22   PUB, "About Water Efficiency Labelling Scheme," https://www.pub.gov.sg/wels/about.

23   Brears, "Urban Water Security in Asia-Pacific: Promoting Demand Management Strategies".

24   Isaac B. Addo, Martin C. Thoms, and Melissa Parsons, "The Influence of Water-Conservation Messages on Reducing Household Water Use," *Applied Water Science* 9, no. 5 (2019).

25   GWP, "Raising Public Awareness," https://www.gwp.org/en/learn/iwrm-toolbox/Management-Instruments/Promoting_Social_Change/Raising_public_awareness/.

26   Georgia Environmental Protection Division Watershed Protection Branch, (2007), https://www1.gadnr.org/cws/Documents/Conservation_Education.pdf

27   Rabab I. El-Nwsany, Ibrahim Maarouf, and Waled Abd el-Aal, "Water Management as a Vital Factor for a Sustainable School," *Alexandria Engineering Journal* 58, no. 1 (2019).

28   Damian C. Adams et al., "The Influence of Water Attitudes, Perceptions, and Learning Preferences on Water-Conserving Actions," *Natural Sciences Education* 42, no. 1 (2013).

29   Scottish Water, "Education," https://www.scottishwater.co.uk/help-and-resources/education.

30   R.C. Brears, *The Green Economy and the Water-Energy-Food Nexus* (London: Palgrave Macmillan UK, 2017).

31   Sydney Water, "Quakers Hill Purified Recycled Water (Prw) Demonstration Plant," https://www.sydneywatertalk.com.au/quakers-hill-prw-demo-plant.

32   Victorian Government Department of Health, "Guidelines for Water Reuse and Recycling in Victorian Health Care Facilities: Non-Drinking Applications," (2009), https://www2.health.vic.gov.au/Api/downloadmedia/%7B949656D2-00DA-486E-B450-84C75D71A0BF%7D.

33   Australian Water Association, "Water Recycling," http://www.awa.asn.au/AWA_MBRR/Publications/Fact_Sheets/Water_Recycling_Fact_Sheet/AWA_MBRR/Publications/Fact_Sheets/Water_Recycling_Fact_Sheet.aspx?hkey=54c6e74b-0985-4d34-8422-fc3f7523aa1d.

34   National Academy of Sciences, "Understanding Water Reuse," (2012), http://dels.nas.edu/resources/static-assets/materials-based-on-reports/booklets/110805697-Understanding-Water-Reuse.pdf

35   Brears, *Developing the Circular Water Economy*.

36   Ibid.

37   Hamburg Wasser, "Hamburg Water Cycle," https://www.hamburgwatercycle.de/en/hamburg-water-cycler/.

38   Hunter Water, "Recycling & Reuse," https://www.hunterwater.com.au/Water-and-Sewer/Recycling–Reuse/.

39   Clemencia Rodriguez et al., "Indirect Potable Reuse: A Sustainable Water Supply Alternative," *International journal of environmental research and public health* 6, no. 3 (2009).

40   Orange County Water District, "Gwrs – New Water You Can Count On," https://www.ocwd.com/gwrs/.

41   Caroline E. Scruggs and Bruce M. Thomson, "Opportunities and Challenges for Direct Potable Water Reuse in Arid Inland Communities," *Journal of Water Resources Planning and Management* 143, no. 10 (2017).

42   J. Lahnsteiner, P. van Rensburg, and J. Esterhuizen, "Direct Potable Reuse – a Feasible Water Management Option," *Journal of Water Reuse and Desalination* 8, no. 1 (2017).

43   American Water Works Association WateReuse, Water Environment Federation, and National Water Research Institute,, "Framework for Direct Potable Reuse" (2015), https://watereuse.org/wp-content/uploads/2015/09/14-20.pdf

**44** Wingoc, "Windhoek Celebrates the 50th Anniversary of Direct Potable Reuse (Dpr) in Namibia," https://www.wingoc.com.na/media/news/windhoek-celebrates-50th-anniversary-direct-potable-reuse-dpr-namibia.

**45** 2030 Water Resources Group, "Wastewater Reclamation to Meet Potable Water Demand: Windhoek, Namibia," (2015), https://www.waterscarcitysolutions.org/wp-content/uploads/2015/08/Wastewater-reclamation-to-meet-potable-water-demand-Windhoek-Namibia.pdf.

# References

2030 Water Resources Group. "Wastewater Reclamation to Meet Potable Water Demand: Windhoek, Namibia". (2015). https://www.waterscarcitysolutions.org/wp-content/uploads/2015/08/Wastewater-reclamation-to-meet-potable-water-demand-Windhoek-Namibia.pdf.

Adams, Damian C., Derek Allen, Tatiana Borisova, Diane E. Boellstorff, Michael D. Smolen, and Robert L. Mahler. "The Influence of Water Attitudes, Perceptions, and Learning Preferences on Water-Conserving Actions". [In English]. *Natural Sciences Education* 42, no. 1 (2013): 114–22.

Addo, Isaac B., Martin C. Thoms, and Melissa Parsons. "The Influence of Water-Conservation Messages on Reducing Household Water Use". *Applied Water Science* 9, no. 5 (2019/06/21 2019): 126.

Anglian Water. "Anglian Water to Trial Fibre-Optic Cables to Find Hidden Leaks in Water Mains". https://www.anglianwater.co.uk/news/anglian-water-to-trial-fibre-optic-cables-to-find-hidden-leaks-in-water-mains/

Australian Goverment. "Water Rating Label". https://www.waterrating.gov.au/choose/water-rating-label

Australian Water Association. "Water Recycling". http://www.awa.asn.au/AWA_MBRR/Publications/Fact_Sheets/Water_Recycling_Fact_Sheet/AWA_MBRR/Publications/Fact_Sheets/Water_Recycling_Fact_Sheet.aspx?hkey=54c6e74b-0985-4d34-8422-fc3f7523aa1d.

Beal, C. D., and J. Flynn. "Toward the Digital Water Age: Survey and Case Studies of Australian Water Utility Smart-Metering Programs". *Utilities Policy* 32 (2015/03/01/ 2015): 29–37.

Berardi, Luigi, Daniele B. Laucelli, Antonietta Simone, Gianfredi Mazzolani, and Orazio Giustolisi. "Active Leakage Control with Wdnetxl". *Procedia Engineering* 154 (2016/01/01/ 2016): 62–70.

Boyle, Thomas, Damien Giurco, Pierre Mukheibir, Ariane Liu, Candice Moy, Stuart White, and Rodney Stewart. "Intelligent Metering for Urban Water: A Review". *Water* 5, no. 3 (2013).

Brears, R.C. *Developing the Circular Water Economy*. Cham, Switzerland: Palgrave Macmillan, 2020.

____. *The Green Economy and the Water-Energy-Food Nexus*. London: Palgrave Macmillan UK, 2017.

____. *Urban Water Security*. Chichester, UK; Hoboken, NJ: John Wiley & Sons, 2016.

____. "Urban Water Security in Asia-Pacific: Promoting Demand Management Strategies". (2014). https://refubium.fu-berlin.de/bitstream/handle/fub188/18349/pp414-urban-water-security-asiapacific.pdf?sequence=1&isAllowed=y.

El-Nwsany, Rabab I., Ibrahim Maarouf, and Waled Abd el-Aal. "Water Management as a Vital Factor for a Sustainable School". *Alexandria Engineering Journal* 58, no. 1 (2019/03/01/ 2019): 303–13.

European Commission. "The Role of Water Pricing and Water Allocation in Agriculture in Delivering Sustainable Water Use in Europe – Final Report" (2012). https://ec.europa.eu/environment/water/quantity/pdf/agriculture_report.pdf.

FAO. "Water Charging in Irrigated Agriculture". http://www.fao.org/3/y5690e/y5690e04.htm

Georgia Environmental Protection Division Watershed Protection Branch. (2007). https://www1.gadnr.org/cws/Documents/Conservation_Education.pdf

Ghosh, S.N. *Environmental Hydrology and Hydraulics: Eco-Technological Practices for Sustainable Development*. CRC Press, 2016.

Güngör, Mahmud, Ufuk Yarar, and Mahmut Firat. "Reduction of Water Losses by Rehabilitation of Water Distribution Network". *Environmental Monitoring and Assessment* 189, no. 10 (2017/09/11 2017): 498.

GWP. "Raising Public Awareness". https://www.gwp.org/en/learn/iwrm-toolbox/Management-Instruments/Promoting_Social_Change/Raising_public_awareness/.

Hamburg Wasser. "Hamburg Water Cycle". https://www.hamburgwatercycle.de/en/hamburg-water-cycler/.

Helix Water District. "Our Demonstration Landscape Is Complete". https://hwd.com/our-demonstration-landscape-is-complete/.

Hunter Water. "Recycling & Reuse". https://www.hunterwater.com.au/Water-and-Sewer/Recycling-Reuse/.

Irish Water. "Your Business Charges". https://www.water.ie/for-business/billing-explained/charges/.

Kelly, D. A. "Labelling and Water Conservation: A European Perspective on a Global Challenge". [In English]. *Building Services Engineering Research & Technology* 36, no. 6 (Nov 2015-11-19 2015): 643–57.

Lahnsteiner, J., P. van Rensburg, and J. Esterhuizen. "Direct Potable Reuse – a Feasible Water Management Option". *Journal of Water Reuse and Desalination* 8, no. 1 (2017): 14–28.

National Academy of Sciences. "Understanding Water Reuse". (2012). http://dels.nas.edu/resources/static-assets/materials-based-on-reports/booklets/110805697-Understanding-Water-Reuse.pdf

New South Wales Government. "Level 2 Water Restrictions to Start across Sydney". https://www.nsw.gov.au/news/level-2-water-restrictions-to-start-across-sydney.

Orange County Water District. "Gwrs – New Water You Can Count On". https://www.ocwd.com/gwrs/.

Rodriguez, Clemencia, Paul Van Buynder, Richard Lugg, Palenque Blair, Brian Devine, Angus Cook, and Philip Weinstein. "Indirect Potable Reuse: A Sustainable Water Supply Alternative". [In eng]. *International journal of environmental research and public health* 6, no. 3 (2009): 1174–209.

San Diego County Water Authority. "New Agricultural Water Program Benefits San Diego County Growers". https://www.sdcwa.org/new-agricultural-water-program-benefits-san-diego-county-growers.

⎯⎯⎯. "Outreach and Education". https://www.sdcwa.org/outreach-and-education.

Scruggs, Caroline E., and Bruce M. Thomson. "Opportunities and Challenges for Direct Potable Water Reuse in Arid Inland Communities". *Journal of Water Resources Planning and Management* 143, no. 10 (2017): 04017064.

State of Green. "Reducing Urban Water Loss". (2016). https://stateofgreen.com/en/publications/reducing-urban-water-loss/.

UNEP. "Source Book of Alternative Technologies for Freshwater Augmentation in Latin America and the Caribbean". https://www.oas.org/dsd/publications/Unit/oea59e/ch20.htm#2.1%20desalination%20by%20reverse%20osmosis.

Victorian Government Department of Health. "Guidelines for Water Reuse and Recycling in Victorian Health Care Facilities: Non-Drinking Applications". (2009). https://www2.health.vic.gov.au/Api/down loadmedia/%7B949656D2-00DA-486E-B450-84C75D71A0BF%7D.

WateReuse, American Water Works Association, Water Environment Federation, and National Water Research Institute,. "Framework for Direct Potable Reuse" (2015). https://watereuse.org/wp-content/uploads/2015/09/14-20.pdf

Wingoc. "Windhoek Celebrates the 50th Anniversary of Direct Potable Reuse (Dpr) in Namibia". https://www.wingoc.com.na/media/news/windhoek-celebrates-50th-anniversary-direct-potable-reuse-dpr-namibia.

Yorkshire Water. "Yorkshire Water Looking to Save Millions of Litres of Water with New Smart Meters". https://www.yorkshirewater.com/news-media/2020/yorkshire-water-looking-to-save-millions-of-litres-of-water-with-new-smart-meters/.

City of Johannesburg. "Level 1 Water Restrictions Remain in Place for Joburg." https://www.joburg.org.za/media_/Newsroom/Pages/2019%20Newsroom%20Articles/April%202019/Level-1-water-restrictions-remain-in-place-in-Joburg.aspx#:~:text=Johannesburg%20Water%2C%20in%20an%20effort%20to%20maintain%20a,·%20Do%20not%20leave%20taps%20dripping.%20More%20items.

iTnews. "Sydney Water Taps Unsw Fibre Optic Sonar to Find Leaks." https://www.itnews.com.au/news/sydney-water-taps-unsw-fibre-optic-sonar-to-find-leaks-525862.

PUB. "About Water Efficiency Labelling Scheme." https://www.pub.gov.sg/wels/about.

Scottish Water. "Education." https://www.scottishwater.co.uk/help-and-resources/education.

Severn Trent. "Severn Trent to Begin New Smart Meter Trial in West Bridgford to Combat Leakage." https://www.stwater.co.uk/news/news-releases/severn-trent-to-begin-new-smart-meter-trial-in-west-bridgford-to/.

SFID. "Permanent Special Agricultural Water Rate Program (Psawr)." https://www.sfidwater.org/288/Permanent-Special-Agricultural-Water-Rat#:~:text=The%20uniform%20rate%20for%20all%20agricultural%20water%20use,the%20regular%20rate%20%28s%29.%20Frequently%20Asked%20Questions%20%28link%29.

Sydney Water. "Quakers Hill Purified Recycled Water (Prw) Demonstration Plant." https://www.sydneywatertalk.com.au/quakers-hill-prw-demo-plant.

Toronto Water. "Industrial Water Rate Program." https://www.toronto.ca/services-payments/water-environment/how-to-use-less-water/water-efficiency-for-business/industrial-water-rate-program/#:~:text=The%20Industrial%20Water%20Rate%20Program,m3)%20of%20water%20annually.

Toronto Water. "Industrial Water Rate Program." https://www.toronto.ca/services-payments/water-environment/how-to-use-less-water/water-efficiency-for-business/industrial-water-rate-program/#:~:text=The%20Industrial%20Water%20Rate%20Program,m3)%20of%20water%20annually.

# Chapter 4
# Generating renewable energy and recovering resources from wastewater

**Abstract:** Wastewater production is expected to increase significantly as the century progresses. This chapter will discuss how wastewater treatment plants should not be viewed as waste disposal facilities, but rather as water resource recovery facilities that produce clean water, reduce dependence on fossil fuels through the use and production of renewable energy, and recover nutrients.

**Keywords:** Wastewater, Water Resource Recovery, Renewable Energy, Nutrient Recovery

## Introduction

Currently, it is estimated that the world produces 380 billion cubic metres of wastewater per annum. Globally, wastewater production is expected to increase by 24 percent by 2030 and 51 percent by 2050. It is estimated that energy embedded in wastewater would be enough to provide electricity to 158 million households. Among major nutrients, 16.6 million metric tonnes (Tg) of nitrogen is estimated to be embedded in wastewater produced per annum while phosphorous stands at 3.0 Tg and potassium at 6.3 Tg. The full nutrient recovery from wastewater would offset 13.4 percent of the global demand for these nutrients in agriculture.[1] This chapter will discuss how wastewater treatment plants should not be viewed as waste disposal facilities, but rather as water resource recovery facilities that produce clean water, reduce dependence on fossil fuels through the use and production of renewable energy, and recover nutrients.[2,3]

## 4.1 Renewable energy generation technologies at wastewater treatment facilities

Electricity is the main energy source required in wastewater treatment plants, accounting for around 25–50 percent of the operating costs of traditional activated sludge plants.[4] Energy derived from wastewater treatment is a renewable energy resource. It can include:

- Electrical energy, heat, or biofuels from the utilisation of digester gas (biogas that consists mainly of methane and carbon dioxide)
- Electrical energy and heat from thermal conversion of biomass (biosolids)
- Heating or cooling energy using plant influent or effluent as a heat source or sink for a heat pump[5]

https://doi.org/10.1515/9783111028101-004

### 4.1.1 Biogas from anaerobic digestion

Anaerobic digestion is a proven technology for sewage sludge treatment and allows the generation of renewable energy from the same process. During anaerobic digestion, microorganisms break down the organic matter contained in the sludge and convert it into biogas which can be used for electricity, heat, and biofuel production. Specifically, the sludge is pumped into anaerobic continuously stirred tank reactors where digestion takes place, usually at mesophilic temperatures (35–39 degrees Celsius). During a retention time, usually around 20 days, microorganisms break down part of the organic matter that is contained in the sludge and produce biogas, which is composed of methane, carbon dioxide, and trace gases. The raw biogas is dried, and hydrogen sulphide and other trace substances removed to obtain a good combustible gas and avoid corrosion or unwanted deposition in the combustion equipment. After cleaning, the biogas can be upgraded to biomethane, or it can be combusted in a combined heat and power (CHP) plant to generate electricity and heat simultaneously.[6]

### 4.1.2 Biomethane

Biogas is primarily composed of methane (40–70 volume percent) and carbon dioxide and smaller traces of acidic gases and impurities such as hydrogen sulphide, nitrogen, water vapour, and traces of other volatile organic gases. Biomethane is produced via biogas upgrading, which is the removal of carbon dioxide before the biogas can be used as a vehicle fuel or injected into the natural gas grid, as the large volume of carbon dioxide reduces its heating value. Biogas can be upgraded using the following technologies:

– *Cryogenic separation*: This involves cooling the acid gases to a very low temperature so that the carbon dioxide can be liquefied and separated
– *Membrane separation*: This technique uses polymeric membranes to separate the carbon dioxide from the methane in biogas while under high pressure
– *Organic physical scrubbing*: Carbon dioxide is more soluble than methane. Raw biogas flows through a counter flow of a liquid in a column. The liquid absorbs the carbon dioxide, leaving biogas with a high content of methane
– *Pressure swing adsorption*: In this process, biogas is compressed to a pressure between 4–10 bar and is fed to a vessel (column) where it is put in contact with a material (adsorbent) that will selectively retain carbon dioxide
– *Amine scrubbing and water wash (or water scrubbing)*: Amine systems and water scrubbing are similar in that they are both 'wet' upgrading systems and involve separating the carbon dioxide from the methane by solubilising the carbon dioxide in a liquid solution while allowing the methane to pass[7,8,9]

**Case 4.1: San Antonio Water System's Unique renewable energy strategy**
San Antonio Water System (SAWS) has implemented a unique renewable energy strategy to harness methane gas generated during the wastewater treatment process. This biogas, which is 60 percent methane, is produced as a byproduct of the anaerobic digestion process from biosolids, with approximately 140,000 tons of biosolids generated annually. To harness the biogas, SAWS partnered with Ameresco, Inc., a national energy company focused on renewable energy. Since 2010, Ameresco has been processing more than 1.5 million standard cubic feet of biogas daily and delivering a minimum of 900,000 cubic feet of natural gas each day to a nearby commercial pipeline for sale on the open market. This amount of biogas is enough to fill seven commercial blimps or 1,250 tanker trucks every day. By harnessing the biogas instead of burning it off using flares, SAWS has achieved significant environmental benefits, including the reduction of 31,261 cars from the road, the planting of 38,736 acres of trees, the reduction of 19,739 tons of carbon dioxide, and the heating of more than 4,689 average-size homes. Furthermore, SAWS receives about $200,000 in annual royalties through the sale of the biogas, which reduces the cost of SAWS operations and helps to keep rates affordable. SAWS' implementation of this unique renewable energy strategy has made it the first city to harness biogas in this manner. Together with its water recycling and composting efforts, this further demonstrates SAWS' commitment to providing sustainable and affordable water services for its growing city.[10]

## 4.1.3 Combined heat and power

CHP is the most prevalent means of utilising biogas. As the process of anaerobic digestion requires some heat, it is suited to CHP. The ratio of heat to power varies depending on the scale and technology but typically 35–40 percent is converted to electricity, 40–45 percent to heat and the balance lost as inefficiencies in the various stages of the process. This equates to over 2 kWh electricity and 2.5 kWh heat per cubic metre, at 60 percent methane.[11] CHP offers a variety of benefits, including:

– *Efficiency*: CHP requires less fuel than separate heat and power generation systems to produce a given energy output. CHP also avoids transmission and distribution losses that occur when electricity travels over power lines from central generating units
– *Reliability*: CHP can provide high-quality electricity and thermal energy to a site regardless of what happens on the power grid, decreasing the impacts of outages and improving power quality for sensitive equipment
– *Environmental*: Because less fuel is burned to produce each unit of energy output, CHP reduces greenhouse gases and other air pollutants
– *Economic*: CHP lowers a facility's energy bill considerably due to its high efficiency, and it can provide a hedge against unstable energy costs[12]

**Case 4.2: Tallinna Vesi's Biogas-Powered Combined Heat and Power Plant**
AS Tallinna Vesi, a wastewater treatment company in Tallinn, Estonia, has signed a contract with Filter Solutions OÜ for the construction of a CHP plant at the wastewater treatment plant in Paljassaare. The CHP will use biogas generated during the stabilisation of sewage sludge to produce electricity and

heat, helping to reduce the cost of electricity and heat while adding value to the company's by-product. The plant is expected to be launched by the end of 2023, with a construction cost of approximately EUR 2.4 million. The CHP plant includes biogas purification and two Jenbacher gas engines, with a biogas consumption capacity of up to 4.6 MW. At this capacity, the electricity generation capacity is 1.9 MW, and the heat generation capacity is up to 2.1 MW, fully used for the wastewater treatment plant's own use. The biogas generated annually from sewage sludge can cover the electricity needs of approximately 3,800 average households. The construction of the CHP is part of a larger project that includes the reconstruction of digesters to maximize the energy potential of sewage sludge. The wastewater treatment plant is equipped with a buffer reservoir that can store up to 20 MWh of biogas, serving as an energy storage facility. The investment will reduce the annual electricity consumption of the wastewater treatment plant and the pumping station by nearly 50%, with biogas covering up to 70% of the plant's electricity needs. The investment supports Tallinna Vesi's goal of reducing its environmental impact, with 100% of the electricity used at the company's facilities and in the treatment process generated from renewable sources. The wastewater treatment plant's renewable energy production volume will exceed the proposals to the Urban Wastewater Treatment Directive published by the European Commission for 2030.[13]

## 4.1.4 Anaerobic co-digestion

In addition to sewage sludge, some wastewater treatment plants include other organic feedstock in the anaerobic reaction. Known as anaerobic co-digestion, it can lead to a significant increase in gas production as most co-substrates have higher methane production per tonne of fresh matter than sewage sludge. This is due to lower water content and high contents of energy-rich substances, including:

- Lipid wastes including fats, oils, and greases (known as FOG)
- Simple carbohydrate wastes, including bakery waste, brewery waste, and sugar-based solutions such as soft drinks
- Complex carbohydrate wastes, including fruits and vegetables as well as mixed organics, including the organic fraction of a municipal solid waste stream
- Protein wastes, including meat, poultry, and dairy waste products
- Other waste organic feedstocks, including glycerine from biosolid fuel production[14,15]

**Case 4.3: A Successful Trial of Co-Digestion of Food Waste and Used Water Sludge in Singapore's Water Treatment Process**

A recent trial project by Singapore's National Water Agency, PUB, and the National Environment Agency (NEA), has demonstrated the potential of co-digestion of food waste and used water sludge to significantly increase biogas yield, supporting Singapore's vision for a Zero Waste Nation and a circular economy. The project, which began in December 2016 and lasted two years, explored the viability of collecting and transporting source-segregated food waste from various premises to a demonstration facility at the Ulu Pandan Water Reclamation Plant for co-digestion with used water sludge. Up to 40 tons of used water sludge and food waste from 23 premises were treated daily at the facility. The mixture of used water sludge and food waste then underwent anaerobic digestion, a biological process that breaks down organic materials in the absence of oxygen, to produce biogas for energy generation. The trial project results showed that co-digestion of food waste and used water sludge can triple biogas yield compared to the treatment of used water sludge alone. Synergistic effects in the co-digestion of used water sludge and food waste can increase biogas production by up to 40% compared to the separate digestion of the two inputs. These results demonstrate that co-digestion of food waste and used water sludge is a viable approach to maximize resource recovery and achieve energy self-sufficiency in the water treatment process. The project's success has led to plans to implement co-digestion of food waste and used water sludge at the new Integrated Waste Management Facility (IWMF) and Tuas Water Reclamation Plant (WRP), collectively known as the Tuas Nexus. These facilities are scheduled to be completed in 2025 and will harness potential synergies and reap the benefits of a Water-Energy-Waste Nexus to maximize both resource and energy recovery while minimizing environmental footprint.[16]

## 4.1.5 Thermal conversion of biosolids

Thermal oxidation (incineration), which is the complete oxidation of organics (biomass) to carbon dioxide and water in the presence of excess air, is a well-established technology. The benefits of thermal conversion include the reduction in biosolids mass, generation of heat for use in heating or electricity generation, reduction in the facility's overall carbon footprint, lowering of the reliance on fossil fuels, generation of ash for use in building materials, and generation of additional revenue to utilities.[17]

### Case 4.4: T-PARK, the World's Largest Water and Energy Autonomous Sewage Treatment Plant in Hong Kong

T-PARK, located in Hong Kong, is the largest sewage treatment plant in the world and is entirely autonomous in terms of water and energy supply, employing advanced technologies and an ecological approach. The plant, designed by Claude Vasconi, was built and is operated by Veolia, and processes sludge from eleven sewage treatment plants in a region with over 7 million inhabitants. The plant's fluidized bed incineration technology reduces waste by 90%, thus reducing the volume of sewage sludge to be disposed of. The heat generated during the process is recovered to generate electricity, producing up to 14 MW, which is sufficient to supply the plant's needs, with the excess being supplied to the public power grid. When running at full capacity, the plant can produce up to 2 MW of surplus electricity, which can light up 4,000 homes. The plant is also equipped with a desalination plant that purifies seawater to meet the site's water needs, and rainwater is also recovered. Furthermore, the plant achieves zero effluent discharge into the sea through a small wastewater treatment plant that collects, processes and reuses the wastewater produced on site. The plant is a multipurpose facility that combines advanced technologies with ecological leisure facilities and educational activities that highlight the benefits of a circular approach to waste management based on energy recovery. The facility includes a recreational and educational center for the general public, a landscaped ecological garden, a theatre, food court, and spa with three heated pools, all of which are heated using the heat recovered during sludge incineration.[18]

### 4.1.6 Thermal energy recovery from wastewater

Thermal energy can be recovered from raw wastewater or effluent by exploiting the significant temperature differential between wastewater and ambient conditions. This temperature difference can be recovered for use in heating and cooling systems, which is often used for buildings at the facility and in buildings of areas surrounding the facility.[19]

#### 4.1.6.1 District heating and cooling

District heating and cooling (DHC) is considered more efficient than the individual, distributed systems for heating and cooling as DHC solutions can utilise locally available, low-cost energy sources. Wastewater heat recovery applications based on heat pumps are becoming more widespread in energy-saving applications for both heating and cooling. Heat recovery can be performed inside the buildings (domestic-scale), from sewerage lines (urban-scale), and from wastewater treatment plants (municipal-scale). In densely populated areas, heat recovery from sewage has immense potential, particularly when Geographic Information System-based analysis is used to match the availability of sewage and heat demand.[20]

> **Case 4.5: Exploring the Viability of a Large Heat Pump Using Seawater and Wastewater for District Heating in Copenhagen**
>
> A group of partners, including three Danish heating companies, are collaborating to establish a large demonstration heat pump utilizing seawater and wastewater as heating sources. The heat pump will be powered by electricity that is expected to be 100% green in Denmark by 2030. The project, called "Big Heat Pumps for District Heating" (SVAF), aims to incorporate more sustainable energy sources into the district heating system of Copenhagen, and ultimately contribute to the city's goal of becoming a fossil-free capital. The new heat pump, located close to the Zealand Bridge, will test the capacity to collect energy and heat from seawater and wastewater, which are viable energy sources for supplying district heating. The project will also explore the reliability, economic effectiveness, and flexibility of the technology. If successful, the solution can increase Denmark's export potential and promote green solutions to other cities.[21]

## 4.2 Renewable energy activities on buildings and surrounding lands

Water utilities can implement renewable energy activities on facility-owned buildings and surrounding lands, including the following.

## 4.2.1 Solar energy

Wastewater treatment plants require many aeration tanks when treating sewage. These require a lot of space in the plant area, providing opportunities to utilise this space with solar photovoltaic (PV) systems to drive equipment or provide heat. These systems, which can produce electricity even in the absence of strong sunlight, can generate significant quantities of electricity depending on a variety of factors including quality of the sunlight and the system's mounted pitch. In addition to lowering energy costs, solar PV systems can improve air quality by reducing pollution caused by using fossil fuels in wastewater treatment plants.[22]

### 4.2.1.1 Floating photovoltaic installations

Floating PV installations are similar to that of land-based PV systems, other than the fact that the PV arrays, and often their inverters, are mounted on a floating platform. They can be installed on reservoirs as well as ponds and lakes. In a floating PV installation, the direct current electricity generated by PV modules is gathered by combiner boxes and converted to alternating current by inverters. For small-scale floating plants close to shore, the inverters can be placed on land a short distance from the array. For larger-scale floating plants, central or string inverters are placed on specially designed floats. The potential benefits of floating solar include:

- Reduced evaporation from water reservoirs, as the solar panels provide shade and limit the evaporative effects of wind
- Improvements in water quality through decreased algae growth
- Reduction or elimination of the shading of panels by their surroundings
- Elimination of the need for major site preparation, such as levelling or the laying of foundations, which are required for land-based installations
- Easy installation and deployment in sites with low anchoring and mooring requirements, with a high level of modularity, leading to faster installations[23]

**Case 4.6: Europe's Largest Floating Solar Park**
Evides, a Dutch water supply company, inaugurated a floating solar park in 2020 on a water reservoir in Rotterdam, the Netherlands. The 1.62 MW installation, developed by Floating Solar, uses 4,787 PV modules aligned with the sun through winches. The size of the park is considered unique in Europe and is equipped with special sensors that measure the sun's strength, wind force, and other parameters to optimize energy production. The moving of the panels is done with winches. The floating solar park covers 30% of the water reservoir, contributing to Evides' sustainability goals and lowering its $CO_2$-footprint. The company plans to monitor the impact of the solar installation on the reservoir, such as algae growth, contamination from bird droppings, reduced UV radiation on the water, and the impact of the wind. Depending on the outcome, Evides may build similar solar parks on three other reservoirs.[24]

## 4.2.2 Wind power

Wind energy, which is captured on-site using wind turbines, can be very cost effective in areas with adequate wind resources. As opposed to large utility-scale wind farm turbines, which can have capacities as high as 3 MW, small wind turbines are often better suited for local facilities. These small wind turbines are most often installed in non-urban areas because installations typically require at least one acre of land and wind speeds averaging around 24 kilometres per hour at 50 metres above the ground.[25]

---

**Case 4.7: Wannon Water's Successful Implementation of Wind Turbine Technology**
Wannon Water, Victoria, Australia, has set a target to reduce greenhouse gas emissions by 40 percent by 2025 and achieve net-zero carbon emissions by 2030. In 2019, Wannon Water constructed an 800-kilowatt wind turbine at the Portland Sewage Treatment Plant, making the water utility the first water corporation in Australia to own and operate a wind turbine of this scale. The project resulted in the first wind turbine 100 percent owned and operated by a Victorian Government entity in more than 30 years. The turbine generates more than two gigawatt hours of renewable energy annually and reduces carbon emissions by an average of 2,500 tonnes per year. This renewable energy powers the city's water and sewage treatment facilities, reducing energy costs and benefiting the environment. The turbine's capital cost of $4.2 million will be paid back within ten years through reduced energy bills. The project supported regional economic growth and prosperity, with local companies sub-contracted to manufacture the tower using local steel and complete the civil works.[26]

---

## 4.2.3 Hydropower energy recovery

Hydropower energy recovery is defined as *"hydropower built using an existing, pressurized, manmade water conveyance that is already diverting water from a natural waterway for the distribution of water for agricultural, municipal, or industrial consumption and not primarily for the generation of electricity"*. Recovering energy from the flow of wastewater entering or leaving a treatment plant using microhydropower turbines is also a viable method of energy savings at plants with large flows rates. Hydropower energy recovery is cost-effective because it is constructed utilising existing infrastructure. The main driver for this type of hydropower is the opportunity for water utilities to lower operational costs by offsetting energy use costs with on-site hydropower generation.[27,28]

---

**Case 4.8: Hydropower Energy Recovery: Converting Aqueduct Energy Dissipation Works into Small Hydropower Plants for Renewable Energy Production in Athens, Greece**
EYDAP, the water utility of Athens, Greece, has initiated a program to utilize renewable energy to optimize energy balance in the country and explore new profitable business opportunities. This program includes hydropower, cogeneration of heat and power, solar energy, and energy reduction initiatives. EYDAP is converting existing energy dissipation works, or small waterfalls along the aqueducts that bring raw water to Athens for treatment, into hydropower plants to take advantage of the aqueducts'

hydropower potential. The process involves temporarily redirecting water through a turbine to generate electricity via a generator before returning the water to the aqueduct at a lower elevation. EYDAP
has constructed six small hydropower plants along its aqueducts in Kirfi, Elikona, Kartala Kihairona,
Mandra, Evinos Dam, and Klidi. Operating small hydropower plants provides numerous benefits, including the exploitation of unused hydraulic energy potential, production of "green" energy using renewable energy sources, protection of the environment, income from the sale of electricity, additional
income from trading carbon rights or green certificates, new employment possibilities, and low operational and maintenance costs due to telemetry and SCADA systems.[29,30]

## 4.3 Energy efficiency

There are various opportunities for improving energy efficiency in wastewater facilities through equipment upgrades (replacing items with more efficient ones), operational modifications (reducing the amount of energy required to perform specific
functions), and modifications to facility buildings (reducing the amount of energy con-
sumed by facility buildings themselves). In the wastewater collection and treatment
process, there are opportunities to increase energy efficiency, including:
- Improving efficiency of aeration equipment and anaerobic digestion
- Implementing cogeneration and other onsite renewable power options
- Implementing lighting, HVAC improvements
- Fixing leaks
- Installing software
- Using efficient pumping systems (pumps, motors, variable frequency drives)
- Recycling water

In the treated wastewater discharge process, there are opportunities to:
- Use efficient pumping systems (pumps, motors, variable frequency drives)
- Capture energy from water moving downhill[31,32]

**Case 4.9: Cape Town Enhancing Energy Efficiency in Wastewater Treatment**
The Cape Town city council has approved three tender contracts for upgrading the Athlone and Bellville Wastewater Treatment Works (WWTW) over the next five years, aimed at improving the energy
efficiency and treated effluent quality of the facilities. The upgrade work includes mechanical and electrical refurbishment of the Athlone WWTW and diffused aeration (DA) reactor upgrades at Bellville
WWTW. The new aeration blowers and diffusers in Athlone WWTW are expected to be more efficient
and use less electricity than the existing aeration system, resulting in increased energy efficiency. The
reconfiguration of the recycle pumps will also ensure optimized nitrogen removal, which has increased
since the treatment plant's inception, contributing to the enhanced treated effluent quality for environmental sustainability. Similarly, the Bellville WWTW DA plant will undergo mechanical and electrical
work and civil work for upgrading the biological reactors to an advanced configuration for enhanced
biological nutrient removal. This will improve the energy efficiency of the DA plant and the treated
effluent quality.[33]

## 4.4 Benefits of renewable energy and energy efficiency

Some of the benefits of implementing renewable energy schemes and improving energy efficiency include:

–   *Reducing air pollution and greenhouse gas emissions*: Increasing the use of renewable energy and improving energy efficiency can help reduce greenhouse gas emissions and air pollutants by decreasing consumption of fossil fuel-based energy. Fossil fuel combustion also generates sulphur dioxide and nitrogen oxide emissions. These pollutants can lead to smog, acid rain, and airborne particulate matter that can cause respiratory health problems for many people

–   *Reducing energy costs*: Local governments can achieve significant cost savings by generating their own electricity and heat from renewable energy systems and increasing their efficiency of wastewater treatment plants

–   *Supporting economic growth through job creation and market development*: Investing in renewable energy systems and energy efficiency can stimulate the local economy and spur development of renewable energy system service and energy efficiency markets. Many of these jobs are performed locally by workers from relatively small local companies as they typically involve installation or maintenance of equipment

–   *Demonstrating leadership*: Investing in renewable energy systems and energy efficiency demonstrates not only responsible government stewardship of tax revenue but also the environmental co-benefits that are obtained from reducing energy usage. The implementation of renewable energy systems and energy efficiency measures may facilitate broader adoption of these technologies and strategies by the private sector

–   *Improving energy and water security*: Improving energy efficiency at wastewater treatment plants reduces electricity demand, avoiding the risk of brownouts or blackouts during high energy demand periods and helping avoid the need to build new power plants, which in turn lowers water requirements to generate electricity

–   *Extending the life of infrastructure/equipment*: Energy-efficient equipment often has a longer service life and requires less maintenance than older, less efficient technologies

–   *Protecting public health*: The deployment of renewable energy systems and improvements in energy efficiency at wastewater treatment plants can reduce air and water pollution from the power plants that supply electricity to those facilities. Equipment upgrades may also allow facilities to increase their capacity for treating wastewater or improving the performance of treatment processes, reducing the risk of waterborne illness[34,35,36]

## 4.5 Recovering resources

Numerous resources can be recovered from wastewater, including the following.

### 4.5.1 Nitrogen and phosphorus

Most wastewaters are relatively diluted, yet their high volumes provide opportunities to recover a sizeable amount of nutrients.[37]

#### 4.5.1.1 Nitrogen

Nitrogenous materials present in the sewage can be removed from sewage effluent and converted into biomass through activated secondary treatment processes. Fertiliser grade ammonium sulphate can be produced from the high ammonia-nitrogen concentration sidestreams from sludge digestion processes by stripping and absorption. The stripping of ammonia can be done by steam (steam is blown through the water, and after condensation, a concentrated ammonia solution is produced) or air (air is bubbled through wastewater and takes up the gaseous ammonia). Zeolites and other minerals such as clay can be used to absorb ammonium.[38,39]

#### 4.5.1.2 Phosphorus

Resource recovery technologies applied to wastewater are generally focused on phosphorus recovery from the biosolids accumulated after the treatment of the main process stream or on sidestreams that have enriched phosphorus because of biological accumulation. The simplest form of beneficial reuse of the recovered phosphorus from sewage treatment is through the land application of biosolids, which can take the form of composted biosolids, alkaline stabilised biosolids, heat-dried pellets, char, or ashes.[40]

#### 4.5.1.3 Struvite recovery

Struvite is mainly known as a scale deposit that naturally occurs under the specific condition of pH and mixing energy in specific areas of wastewater treatment plants, for example, pipes and heat exchanges, when concentrations of magnesium phosphate and ammonia approach an equimolar ratio 1:1:1. However, rather than struvite being a concern to wastewater treatment plants (pipeline blockages and higher plant-wide nutrient load), it can be recovered to reduce phosphorus levels in effluents while simultaneously generating a valuable by-product such as a slow-release fertiliser or raw material for the chemical industry.[41,42]

**Case 4.10: The Metropolitan Water Reclamation District of Greater Chicago and Ostara Nutrient Recovery Technologies Partnership**

The Metropolitan Water Reclamation District of Greater Chicago (MWRD) has partnered with Ostara Nutrient Recovery Technologies to open the world's largest nutrient recovery facility at their Stickney Water Reclamation Plant in Cicero, Ill. The facility will recover phosphorus and nitrogen to create a high-value fertilizer, using Ostara's technology, which is both economically and environmentally viable. The plant will greatly reduce its nutrient effluent load to the Chicago/Calumet river system, upstream of the Mississippi river basin, and will reduce its impact on hypoxia in the Gulf of Mexico. Facing more stringent regulatory limits affecting effluent discharge permits, MWRD sought a closed-loop and cost-effective phosphorus management strategy. Ostara provided MWRD with a solution to their challenges. Ostara's Pearl process for nutrient recovery is based on a closed-loop solution where nutrients such as phosphorus and nitrogen in wastewater are recovered to form a high-value fertilizer that generates revenues for wastewater treatment facilities while helping meet environmental regulations. By recovering nutrients from the treatment facility's wastewater stream and converting them to slow-release fertilizer, Ostara's technology helps MWRD increase operational efficiencies by avoiding struvite build-up and protecting the local watershed. The Pearl process can recover more than 85% of the phosphorus and up to 15% of the nitrogen from wastewater streams before they accumulate as struvite in pipes and equipment. The facility is expected to create annual cost savings in chemicals, solid waste disposal, maintenance, and power. Following the successful commercial start-up of this facility, MWRD plans to implement WASSTRIP, a process that turbo-charges the nutrient recovery process and increases the amount of phosphorus recovered by more than 60%. As a result, the efficiency of the Pearl process will be further enhanced to decrease the magnitude of struvite scale formation and alleviate operational issues.[43]

## 4.5.2 Cellulose

Toilet paper often ends as fibrous particles in the wastewater treatment plant. By using fine-mesh sieves, the cellulose fibres can be successfully removed. The cellulose materials that are recovered can be used:
– To dewater the wastewater treatment plant sewage sludge
– In the production of asphalt
– As a raw material for insulation material[44]

**Case 4.11: The cellulose recovery initiative in the Netherlands**

The Scale-up of low-carbon footprint material recovery techniques in existing wastewater treatment plants (SMART-Plant) project in Europe is aiming to prove the feasibility of circular management of urban wastewater and environmental sustainability of the systems and co-benefits of scaling-up water solutions through Life Cycle Assessment and Life Cycle Costing approaches. The project's cellulose recovery initiative consists of innovative integration of dynamic fine-sieving and in-situ post-processing that has been developed and is currently validated in the municipal wastewater treatment plant of Geestmerambacht in the Netherlands. Filter systems have been installed, separating cellulose fibres from toilet paper in the wastewater. The result is marketable cellulose that has been cleaned, dried, and disinfected while the sludge is sent for post-processing inside the treatment plant.[45]

### 4.5.3 Bioplastic

One of the most non-traditional technologies under development is the production of biodegradable plastic using polymers isolated from biosolids. Polymers contain carbon, hydrogen, oxygen, and nitrogen, and therefore biological wastewater can be used to make polymers. Polymers called polyhydroxyalkanoates (PHA) can be produced by anaerobic bacteria by metabolising renewable organic carbon sources. PHA polymers are biodegradable thermoplastics and can be used as a substitute for conventional petroleum-based plastics.[46]

**Case 4.12: World's first kilogram of PHA from wastewater in the Netherlands**
In 2015, three Dutch water authorities, Brabantse Delta, De Dommel, and Wetterskip Fryslân, in collaboration with STOWA (Dutch Foundation for Applied Water Research), sludge treatment plant SNB, and two commercial parties, Veolia and KNN, produced the world's first kilogram of PHA. It was produced by bacteria from a wastewater treatment facility in the Dutch province of Zeeland. While the capacity of the project is small – a few kilograms a week – the aim is to scale this up to include the total treated wastewater volume and ultimately result in production capacity of 2,000 metric tonnes/year.[47]

### 4.5.4 Bricks and tiles

Sewage sludge ash is the by-product from the combustion of dewatered sewage sludge in an incinerator. The ash is primarily a silty material with some sand-sized particles. The size range and properties of the ash depend on the type of incineration system and the chemical additives used in the wastewater treatment process. The ash can be used in the brick and tile industry.[48]

**Case 4.13: Thames Water helping create energy-efficient bricks from sewage ash**
Thames Water has signed a deal with a private-sector contractor to create energy-efficient bricks from sewage ash. Each day, wastewater enters Europe's largest sewage works in Beckton with the leftover solids used in the utility's waste-to-energy incinerator. Until now the leftover ash was disposed of in a landfill. Thames Water will now provide a contractor with the dried residue ash needed to create the bricks with the ash to be reacted and mixed with carbon dioxide, water, sand, and a small quantity of cement to form aggregate for 17-kilogram blocks. Overall, Thames Water will supply ash to make 18,000 tonnes of the aggregate, enough to create 2.3 million heavy-duty bricks.[49]

### 4.5.5 Mining wastewater for metals

Metals can be potentially mined from wastewater, for instance, silver and cadmium is increasingly being found in wastewater and is expensive enough to potentially warrant recovery.[50]

---

**Case 4.14: The ZERO BRINE project in the Netherlands**

The ZERO BRINE project in the Netherlands aims to prove that minerals can be recovered from industrial processes for reuse in other industries. The Demineralized Water Plant, in the Botlek area owned by EVIDES, is a large-scale demonstration of the ZERO BRINE project that uses a combination of ion exchanges and membrane technology: dissolved air flotation, reverse osmosis, and mixed bed ion exchange. Waste heat and wastewater streams will be combined in a multi-company site environment:
– Eliminating brine effluent (target: zero liquid discharge) of the industrial water supplier
– Recovering high purity magnesium products (target: magnesium purity >90 percent), sodium chloride solution, and sulphate salts
– Recycling streams within the site (target: >70 percent internal recycling of materials recovered)[51]

---

## Notes

**1** Manzoor Qadir et al., "Global and Regional Potential of Wastewater as a Water, Nutrient and Energy Source," *Natural Resources Forum* n/a, no. n/a (2020).

**2** R.C. Brears, *Developing the Circular Water Economy* (Cham, Switzerland: Palgrave Macmillan, 2020).

**3** Ranjani B. Theregowda et al., "Nutrient Recovery from Municipal Wastewater for Sustainable Food Production Systems: An Alternative to Traditional Fertilizers," *Environmental engineering science* 36, no. 7 (2019).

**4** Marta Gandiglio et al., "Enhancing the Energy Efficiency of Wastewater Treatment Plants through Co-Digestion and Fuel Cell Systems," *Frontiers in Environmental Science* 5, no. 70 (2017).

**5** Brears, *Developing the Circular Water Economy*.

**6** IEA Bioenergy, "Sustainable Biogas Production in Municipal Wastewater Treatment Plants," (2015), https://www.ieabioenergy.com/publications/sustainable-biogas-production-in-municipal-wastewater-treatment-plants/.

**7** Ahmed M. I. Yousef et al., "Upgrading Biogas by a Low-Temperature Co2 Removal Technique," *Alexandria Engineering Journal* 55, no. 2 (2016).

**8** Amir I. Adnan et al., "Technologies for Biogas Upgrading to Biomethane: A Review," *Bioengineering* 6, no. 4 (2019).

**9** Saeid Mokhatab, William A. Poe, and John Y. Mak, "Chapter 7 – Natural Gas Treating," in *Handbook of Natural Gas Transmission and Processing (Fourth Edition)*, ed. Saeid Mokhatab, William A. Poe, and John Y. Mak (Gulf Professional Publishing, 2019).

**10** San Antonio Water System, "Biogas," https://www.saws.org/your-water/water-recycling/biogas/.

**11** NNFCC Biocentre, "Biogas," http://www.biogas-info.co.uk/about/biogas/.

**12** US EPA, "Opportunities for Combined Heat and Power at Wastewater Treatment Facilities: Market Analysis and Lessons from the Field," (2011), https://www.epa.gov/sites/production/files/2015-07/documents/opportunities_for_combined_heat_and_power_at_wastewater_treatment_facilities_market_analysis_and_lessons_from_the_field.pdf.

**13** GlobalNewsWire, "As Tallinna Vesi Signed a Contract for the Construction of a Combined Heat and Power Plant on the Wastewater Treatment Plant in Paljassaare," https://www.globenewswire.com/news-release/2023/01/17/2589815/0/en/AS-Tallinna-Vesi-signed-a-contract-for-the-construction-of-a-combined-heat-and-power-plant-on-the-Wastewater-Treatment-Plant-in-Paljassaare.html.

**14** IEA Bioenergy, "Sustainable Biogas Production in Municipal Wastewater Treatment Plants".

**15** US EPA, "Food Waste to Energy: How Six Water Resource Recovery Facilities Are Boosting Biogas Production and the Bottom Line" (2014), https://www.epa.gov/sites/production/files/2016-07/documents/food_waste_to_energy_-_final.pdf.

**16** NEA, "Co-Digestion of Food Waste and Used Water Sludge Enhances Biogas Production for Greater Energy Generation," https://www.nea.gov.sg/media/news/news/index/co-digestion-of-food-waste-and-used-water-sludge-enhances-biogas-production-for-greater-energy-generation.

**17** National Biosolids Partnership, "The Potential Power of Renewable Energy Generation from Wastewater and Biosolids Fact Sheet," (2014), https://www.resourcerecoverydata.org/Potential_Power_of_Renewable_Energy_Generation_From_Wastewater_and_Biosolids_Fact_Sheet.pdf.

**18** Veolia, "The Sewage Sludge Treatment Plant of the Future Is in Hong Kong," https://www.up-to-us.veolia.com/en/water/autonomous-sewage-sludge-treatment-plant-hong-kong-T-PARK.

**19** National Biosolids Partnership, "The Potential Power of Renewable Energy Generation from Wastewater and Biosolids Fact Sheet".

**20** M.; Scoccia Aprile, R.; Dénarié, A.; Kiss, P.; Dombrovszky, M.; Gwerder, D.; Schuetz, P.; Elguezabal, P.; Arregi, B., "District Power-to-Heat/Cool Complemented by Sewage Heat Recovery," *Energies* 12, no. 3 (2019).

**21** DBDH, "Wastewater and Seawater Tested as Sources for Copenhagen District Heating," https://dbdh.dk/wastewater-and-seawater-tested-as-sources-for-copenhagen-district-heating/.

**22** Ziyang Guo et al., "Integration of Green Energy and Advanced Energy-Efficient Technologies for Municipal Wastewater Treatment Plants," *International journal of environmental research and public health* 16, no. 7 (2019).

**23** World Bank, "Where Sun Meets Water: Floating Solar Market Report," (2019), http://documents.worldbank.org/curated/en/579941540407455831/Floating-Solar-Market-Report-Executive-Summary.

**24** Dutch Water Sector, "Rotating Floating Solar Park for Water Utility Evides," https://www.dutchwatersector.com/news/rotating-floating-solar-park-for-water-utility-evides.

**25** US EPA, "On-Site Renewable Energy Generation: A Guide to Developing and Implementing Greenhouse Gas Reduction Programs," (2014), https://www.energy.gov/sites/prod/files/2018/11/f57/onsiterenewables508.pdf.

**26** Wannon Water, "Our Commitment to Combat Climate Change," https://www.wannonwater.com.au/stronger-communities/caring-for-our-environment/our-commitment-to-combat-climate-change.aspx.

**27** National Renewable Energy Laboratory, "Energy Recovery Hydropower: Prospects for Off-Setting Electricity Costs for Agricultural, Municipal, and Industrial Water Providers and Users. July 2017 – September 2017," (2018), https://www.nrel.gov/docs/fy18osti/70483.pdf.

**28** Christine Power, Paul Coughlan, and Aonghus McNabola, "Microhydropower Energy Recovery at Wastewater-Treatment Plants: Turbine Selection and Optimization," *Journal of Energy Engineering* 143, no. 1 (2017).

**29** EYDAP, "Hydropower Projects," https://www.eydap.gr/en/TheCompany/Energy/HydroProjects/.

**30** Robert C. Brears, "The Water-Energy Nexus in Athens," Mark and Focus, https://medium.com/mark-and-focus/the-water-energy-nexus-in-athens-4c496ac778bb.

**31** US EPA, "Energy Efficiency in Water and Wastewater Facilities: A Guide to Developing and Implementing Greenhouse Gas Reduction Programs," (2013), https://www.epa.gov/sites/production/files/2015-08/documents/wastewater-guide.pdf.

**32** Gandiglio et al., "Enhancing the Energy Efficiency of Wastewater Treatment Plants through Co-Digestion and Fuel Cell."

**33** Cape Business News, "R460m Boost for Treated Effluent Quality and Energy-Efficiency at Athlone and Bellville Wastewater Plants," https://www.cbn.co.za/industry-news/sustainability/r460m-boost-for-treated-effluent-quality-and-energy-efficiency-at-athlone-and-bellville-wastewater-plants/.

**34** IEA Bioenergy, "Sustainable Biogas Production in Municipal Wastewater Treatment Plants".

**35** Ibrahim A. Nassar, Kholoud Hossam, and Mahmoud Mohamed Abdella, "Economic and Environmental Benefits of Increasing the Renewable Energy Sources in the Power System," *Energy Reports* 5 (2019).

**36** Jonathan J. Buonocore et al., "Climate and Health Benefits of Increasing Renewable Energy Deployment in the United States," *Environmental Research Letters* 14, no. 11 (2019).

**37** Ka Leung Lam, Ljiljana Zlatanović, and Jan Peter van der Hoek, "Life Cycle Assessment of Nutrient Recycling from Wastewater: A Critical Review," *Water Research* 173 (2020).

**38** IWA, "State of the Art Compendium Report on Resource Recovery from Water," (2016), http://www.iwa-network.org/publications/state-of-the-art-compendium-report-on-resource-recovery-from-water/.

**39** Jianyin Huang et al., "Removing Ammonium from Water and Wastewater Using Cost-Effective Adsorbents: A Review," *Journal of Environmental Sciences* 63 (2018).

**40** Stewart Burn, Tim Muster, and Anna Kaksonen, *Resource Recovery from Wastewater: A Research Agenda. Werf Research Report Series* (London, UNITED KINGDOM: IWA Publishing, 2014).

**41** K. S. Le Corre et al., "Phosphorus Recovery from Wastewater by Struvite Crystallization: A Review," *Critical Reviews in Environmental Science and Technology* 39, no. 6 (2009).

**42** Bing Li et al., "Phosphorous Recovery through Struvite Crystallization: Challenges for Future Design," *Science of The Total Environment* 648 (2019).

**43** Wastewater Digest, "Mwrd, Ostara Open World's Largest Nutrient Recovery Facility," https://www.wwdmag.com/home/news/10934279/mwrd-ostara-open-worlds-largest-nutrient-recovery-facility.

**44** Sabine; Mulder Eijlander, Karel F, "Sanitary Systems: Challenges for Innovation," *Journal of Sustainable Development of Energy, Water and Environment Systems* 7, no. 2 (2019).

**45** SMART-Plant, "Scale-up of Low-Carbon Footprint Material Recovery Techniques in Existing Wastewater Treatment Plants," https://www.smart-plant.eu/index.php/cellulosecellulose-recovery.

**46** Polymer Solutions, "Wastewater Put to Use Making Bioplastics," https://www.polymersolutions.com/blog/wastewater-put-to-use-making-bioplastics/.

**47** Bioplastics Magazine, "World First – Pha from Sewage Sludge," https://www.bioplasticsmagazine.com/en/news/meldungen/20151023-Sewage-based-PHA-produced.php.

**48** Brears, *Developing the Circular Water Economy*.

**49** New Civil Engineer, "Millions of Bricks to Be Made of Recycled Sewage Waste," https://www.newcivilengineer.com/latest/millions-of-bricks-to-be-made-of-recycled-sewage-waste/10041573.article.

**50** Brears, *Developing the Circular Water Economy*.

**51** ZERO BRINE, "Water Plant I Netherlands," https://zerobrine.eu/pilot-projects/netherlands/.

# References

Adnan, Amir I., Mei Y. Ong, Saifuddin Nomanbhay, Kit W. Chew, and Pau L. Show. "Technologies for Biogas Upgrading to Biomethane: A Review". *Bioengineering* 6, no. 4 (2019).

Aprile, M.; Scoccia, R.; Dénarié, A.; Kiss, P.; Dombrovszky, M.; Gwerder, D.; Schuetz, P.; Elguezabal, P.; Arregi, B., "District Power-to-Heat/Cool Complemented by Sewage Heat Recovery". *Energies* 12, no. 3 (2019).

Aquatech. "Europe's Largest Wastewater Plant Saves £500k/Year". https://www.aquatechtrade.com/news/wastewater/thames-water-wastewater-treatment/.

Bioplastics Magazine. "World First – Pha from Sewage Sludge". https://www.bioplasticsmagazine.com/en/news/meldungen/20151023-Sewage-based-PHA-produced.php.

Brears, R.C. *Developing the Circular Water Economy*. Cham, Switzerland: Palgrave Macmillan, 2020.

____. "A Hybrid Sewage Power Plant". Mark and Focus, https://medium.com/mark-and-focus/a-hybrid-sewage-power-plant-3e86805df310

____. "Stockholm Turning Wastewater into Resourcewater". Mark and Focus, https://medium.com/mark-and-focus/stockholm-turning-wastewater-into-resourcewater-6bf27e8028e5

Buonocore, Jonathan J., Ethan J. Hughes, Drew R. Michanowicz, Jinhyok Heo, Joseph G. Allen, and Augusta Williams. "Climate and Health Benefits of Increasing Renewable Energy Deployment in the United States". *Environmental Research Letters* 14, no. 11 (2019/10/29 2019): 114010.

Burn, Stewart, Tim Muster, and Anna Kaksonen. *Resource Recovery from Wastewater: A Research Agenda. Werf Research Report Series*. London, UNITED KINGDOM: IWA Publishing, 2014.

CHP Technical Assistance Partnerships. "Mcalpine Creek Wastewater Management Facility 1mw Biogas Chp System". (2019). http://www.chptap.org/Data/projects/McAlpineWWTP-Project_Profile.pdf.

Eijlander, Sabine; Mulder, Karel F. "Sanitary Systems: Challenges for Innovation". *Journal of Sustainable Development of Energy, Water and Environment Systems* 7, no. 2 (2019): 193–212.

Environmental Protection Department. "Food Waste/Sewage Sludge Anaerobic Co-Digestion Trial Scheme". https://www.epd.gov.hk/epd/english/environmentinhk/waste/prob_solutions/codigestion_trial_scheme.html.

Gandiglio, Marta, Andrea Lanzini, Alicia Soto, Pierluigi Leone, and Massimo Santarelli. "Enhancing the Energy Efficiency of Wastewater Treatment Plants through Co-Digestion and Fuel Cell Systems". [In English]. *Frontiers in Environmental Science* 5, no. 70 (2017-October-30 2017).

Guo, Ziyang, Yongjun Sun, Shu-Yuan Pan, and Pen-Chi Chiang. "Integration of Green Energy and Advanced Energy-Efficient Technologies for Municipal Wastewater Treatment Plants". [In eng]. *International journal of environmental research and public health* 16, no. 7 (2019): 1282.

Huang, Jianyin, Nadeeka Rathnayake Kankanamge, Christopher Chow, David T. Welsh, Tianling Li, and Peter R. Teasdale. "Removing Ammonium from Water and Wastewater Using Cost-Effective Adsorbents: A Review". *Journal of Environmental Sciences* 63 (2018/01/01/ 2018): 174–97.

Hunter Water. "Options on Energy-from-Waste Studied at Hunter Water". https://yourvoice.hunterwater.com.au/sustainable-wastewater/news_feed/options-on-energy-from-waste-studied-at-hunter-water.

IEA Bioenergy. "Sustainable Biogas Production in Municipal Wastewater Treatment Plants". (2015). https://www.ieabioenergy.com/publications/sustainable-biogas-production-in-municipal-wastewater-treatment-plants/.

IWA. "State of the Art Compendium Report on Resource Recovery from Water". (2016). http://www.iwa-network.org/publications/state-of-the-art-compendium-report-on-resource-recovery-from-water/.

Lam, Ka Leung, Ljiljana Zlatanović, and Jan Peter van der Hoek. "Life Cycle Assessment of Nutrient Recycling from Wastewater: A Critical Review". *Water Research* 173 (2020/04/15/ 2020): 115519.

Le Corre, K. S., E. Valsami-Jones, P. Hobbs, and S. A. Parsons. "Phosphorus Recovery from Wastewater by Struvite Crystallization: A Review". *Critical Reviews in Environmental Science and Technology* 39, no. 6 (2009/06/01 2009): 433–77.

Li, Bing, Irina Boiarkina, Wei Yu, Hai Ming Huang, Tajammal Munir, Guang Qian Wang, and Brent R. Young. "Phosphorous Recovery through Struvite Crystallization: Challenges for Future Design". *Science of The Total Environment* 648 (2019/01/15/ 2019): 1244–56.

Mokhatab, Saeid,William A. Poe, and John Y. Mak. "Chapter 7 – Natural Gas Treating". In *Handbook of Natural Gas Transmission and Processing (Fourth Edition)*, edited by Saeid Mokhatab, William A. Poe and John Y. Mak, 231–69: Gulf Professional Publishing, 2019.

Nassar, Ibrahim A., Kholoud Hossam, and Mahmoud Mohamed Abdella. "Economic and Environmental Benefits of Increasing the Renewable Energy Sources in the Power System". *Energy Reports* 5 (2019/11/01/ 2019): 1082–88.

National Biosolids Partnership. "The Potential Power of Renewable Energy Generation from Wastewater and Biosolids Fact Sheet". (2014). https://www.resourcerecoverydata.org/Potential_Power_of_Renew able_Energy_Generation_From_Wastewater_and_Biosolids_Fact_Sheet.pdf.

National Renewable Energy Laboratory. "Energy Recovery Hydropower: Prospects for Off-Setting Electricity Costs for Agricultural, Municipal, and Industrial Water Providers and Users. July 2017 – September 2017". (2018). https://www.nrel.gov/docs/fy18osti/70483.pdf.

New Civil Engineer. "Millions of Bricks to Be Made of Recycled Sewage Waste". https://www.newcivilengi neer.com/latest/millions-of-bricks-to-be-made-of-recycled-sewage-waste/10041573.article.

NNFCC Biocentre. "Biogas". http://www.biogas-info.co.uk/about/biogas/.

Polymer Solutions. "Wastewater Put to Use Making Bioplastics". https://www.polymersolutions.com/blog/ wastewater-put-to-use-making-bioplastics/.

Power, Christine, Paul Coughlan, and Aonghus McNabola. "Microhydropower Energy Recovery at Wastewater-Treatment Plants: Turbine Selection and Optimization". *Journal of Energy Engineering* 143, no. 1 (2017): 04016036.

Qadir, Manzoor, Pay Drechsel, Blanca Jiménez Cisneros, Younggy Kim, Amit Pramanik, Praem Mehta, and Oluwabusola Olaniyan. "Global and Regional Potential of Wastewater as a Water, Nutrient and Energy Source". *Natural Resources Forum* n/a, no. n/a (2020/01/27 2020).

Scottish Water Horizons. "Low Carbon Heat. Naturally". (2019). https://www.scottishwaterhorizons.co.uk/ wp-content/uploads/2019/08/Low-carbon-heat-brochure.pdf

SMART-Plant. "Scale-up of Low-Carbon Footprint Material Recovery Techniques in Existing Wastewater Treatment Plants". https://www.smart-plant.eu/index.php/cellulose-recovery.

Stirling Council. "Stirling Council and Scottish Water Horizons Join Forces for Pioneering Renewables Project". https://www.stirling.gov.uk/news/2018/december-2018/stirling-council-and-scottish-water-horizons-join-forces-for-pioneering-renewables-project/

Sydney Water. "Innovation & Renewable Energy". https://www.sydneywater.com.au/sw/education/water-management/innovationrenewableenergy/index.htm.

The Sydney Morning Herald. "First Sewage-Powered Hydro-Electric Plant in Australia". https://www.smh. com.au/environment/sustainability/first-sewagepowered-hydroelectric-plant-in-australia-20100429-tvd3.html.

Theregowda, Ranjani B., Alejandra M. González-Mejía, Xin Cissy Ma, and Jay Garland. "Nutrient Recovery from Municipal Wastewater for Sustainable Food Production Systems: An Alternative to Traditional Fertilizers". [In eng]. *Environmental engineering science* 36, no. 7 (2019): 833–42.

US EPA. "Energy Efficiency in Water and Wastewater Facilities: A Guide to Developing and Implementing Greenhouse Gas Reduction Programs". (2013). https://www.epa.gov/sites/production/files/2015-08/documents/wastewater-guide.pdf.

——. "Food Waste to Energy: How Six Water Resource Recovery Facilities Are Boosting Biogas Production and the Bottom Line" (2014). https://www.epa.gov/sites/production/files/2016-07/documents/food_waste_to_energy_-_final.pdf.

——. "On-Site Renewable Energy Generation: A Guide to Developing and Implementing Greenhouse Gas Reduction Programs". (2014). https://www.energy.gov/sites/prod/files/2018/11/f57/onsiterenew ables508.pdf.

——. "Opportunities for Combined Heat and Power at Wastewater Treatment Facilities: Market Analysis and Lessons from the Field". (2011). https://www.epa.gov/sites/production/files/2015-07/documents/opportunities_for_combined_heat_and_power_at_wastewater_treatment_facilities_market_analysis_and_lessons_from_the_field.pdf.

Watercare. "New Zealand's First Floating Solar Array Unveiled". https://www.watercare.co.nz/About-us/News-media/New-Zealand%E2%80%99s-first-floating-solar-array-unveiled.

Welsh Water. "Biosolids – Services to Agriculture". https://www.dwrcymru.com/en/My-Wastewater/Biosol ids.aspx.

World Bank. "Where Sun Meets Water: Floating Solar Market Report". (2019). http://documents.world bank.org/curated/en/579941540407455831/Floating-Solar-Market-Report-Executive-Summary.

Yousef, Ahmed M. I., Yehia A. Eldrainy, Wael M. El-Maghlany, and Abdelhamid Attia. "Upgrading Biogas by a Low-Temperature Co2 Removal Technique". *Alexandria Engineering Journal* 55, no. 2 (2016/06/01/ 2016): 1143–50.

ZERO BRINE. "Water Plant I Netherlands". https://zerobrine.eu/pilot-projects/netherlands/.

Cape Business News. "R460m Boost for Treated Effluent Quality and Energy-Efficiency at Athlone and Bellville Wastewater Plants." https://www.cbn.co.za/industry-news/sustainability/r460m-boost-for-treated-effluent-quality-and-energy-efficiency-at-athlone-and-bellville-wastewater-plants/.

DBDH. "Wastewater and Seawater Tested as Sources for Copenhagen District Heating." https://dbdh.dk/ wastewater-and-seawater-tested-as-sources-for-copenhagen-district-heating/.

Dutch Water Sector. "Rotating Floating Solar Park for Water Utility Evides." https://www.dutchwatersector. com/news/rotating-floating-solar-park-for-water-utility-evides.

EYDAP. "Hydropower ProjectsS." https://www.eydap.gr/en/TheCompany/Energy/HydroProjects/.

GlobalNewsWire. "As Tallinna Vesi Signed a Contract for the Construction of a Combined Heat and Power Plant on the Wastewater Treatment Plant in Paljassaare." https://www.globenewswire.com/news-release/2023/01/17/2589815/0/en/AS-Tallinna-Vesi-signed-a-contract-for-the-construction-of-a-combined-heat-and-power-plant-on-the-Wastewater-Treatment-Plant-in-Paljassaare.html.

NEA. "Co-Digestion of Food Waste and Used Water Sludge Enhances Biogas Production for Greater Energy Generation." https://www.nea.gov.sg/media/news/news/index/co-digestion-of-food-waste-and-used-water-sludge-enhances-biogas-production-for-greater-energy-generation.

Robert C. Brears. "The Water-Energy Nexus in Athens." Mark and Focus, https://medium.com/mark-and-focus/the-water-energy-nexus-in-athens-4c496ac778bb.

San Antonio Water System. "Biogas." https://www.saws.org/your-water/water-recycling/biogas/.

Veolia. "The Sewage Sludge Treatment Plant of the Future Is in Hong Kong." https://www.up-to-us.veolia. com/en/water/autonomous-sewage-sludge-treatment-plant-hong-kong-T-PARK.

Wannon Water. "Our Commitment to Combat Climate Change." https://www.wannonwater.com.au/stron ger-communities/caring-for-our-environment/our-commitment-to-combat-climate-change.aspx.

Wastewater Digest. "Mwrd, Ostara Open World's Largest Nutrient Recovery Facility." https://www. wwdmag.com/home/news/10934279/mwrd-ostara-open-worlds-largest-nutrient-recovery-facility.

Veolia. "The Sewage Sludge Treatment Plant of the Future Is in Hong Kong." https://www.up-to-us.veolia. com/en/water/autonomous-sewage-sludge-treatment-plant-hong-kong-T-PARK.

Wannon Water. "Our Commitment to Combat Climate Change." https://www.wannonwater.com.au/stron ger-communities/caring-for-our-environment/our-commitment-to-combat-climate-change.aspx.

Wastewater Digest. "Mwrd, Ostara Open World's Largest Nutrient Recovery Facility." https://www. wwdmag.com/home/news/10934279/mwrd-ostara-open-worlds-largest-nutrient-recovery-facility.

# Chapter 5
# Greening of grey water infrastructure

**Abstract:** Traditionally, stormwater systems are designed to remove stormwater from sites as quickly as possible to reduce on-site flooding. Many cities have implemented these systems as part of a larger sewer system that also regulates domestic and industrial wastewater. The downside of this traditional system is increased peak flows and total discharges from storm events, enhanced delivery of nutrients, and combined sewer overflows during wet conditions. In contrast, various green infrastructure solutions are available to manage stormwater while utilising natural processes to improve water quality.

**Keywords:** Stormwater, Green Infrastructure, Green Roofs, Blue Roofs, Green Streets

## Introduction

Traditionally, stormwater systems, which comprise of stormwater drainpipes, kerb inlets, access holes, culverts and so forth, are designed to remove stormwater from sites as quickly as possible to a main river channel or nearest body of water to reduce on-site flooding. Many cities have implemented these systems as part of a larger sewer system that also regulates domestic and industrial wastewater. Combined sewer systems are the most common where stormwater and wastewater is collected in one pipe network, with the mixed water transported to a wastewater treatment plant for cleaning before being discharged into a river or large body of water. The downside of this traditional system is increased peak flows and total discharges from storm events, enhanced delivery of nutrients degrading aquatic habitats in waterways, and combined sewer overflows during wet conditions. Also, sealed surfaces in urban areas are increasing peak flows, increasing downstream flooding risks and lowering groundwater recharge rates.[1] This chapter will discuss the various green infrastructure solutions available to manage stormwater while utilising natural processes to improve water quality.

## 5.1 Rainwater harvesting

Rainwater harvesting systems comprise three elements: a collection area, a conveyance system, and storage facilities. The collection area is typically the roof of a house or building. A conveyance system usually consists of gutter systems or pipes that deliver rainfall falling on the rooftop to cisterns or other storage vessels. It is recommended that both drainpipes and roof surfaces be constructed of chemically inert

https://doi.org/10.1515/9783111028101-005

materials such as wood, plastic, aluminium, or fibreglass to avoid adverse effects. The water is stored in a storage tank or cistern, which should be constructed of inert material such as reinforced concrete, fibreglass, or stainless steel. The storage tank can be constructed as part of the building or built as a separate unit to the building.

Rainwater is considered naturally clean; however, the collecting surface always introduces contaminants such as sediments, pathogens, metals, organic matter, and volatile organic compounds. The quality of harvested rainwater also depends on the surrounding environment as well as the level of maintenance of the system and storage time. As such, rainwater is usually harvested for a variety of non-potable needs such as irrigation, laundry, and toilet flushing. The following guidelines should be considered when maintaining rainwater harvesting systems:

– A procedure to eliminate 'foul flush' after a long period of dry weather should be considered to remove undesirable materials that have accumulated on the roof and other surfaces between rainfalls
– The storage tank should be checked and cleaned periodically with chlorine solution recommended for cleaning, followed by thorough rinsing
– Care should be taken to keep rainfall collection surfaces covered to reduce the likelihood of mosquitoes using the cistern as a breeding ground
– Gutters and downpipes should be periodically inspected and cleaned
– Community systems require the creation of a community organisation to maintain them effectively[2,3,4,5]

Rainwater harvesting systems are suitable in all areas as a means of augmenting the amount of water available. The main benefits are:

– Rainwater harvesting provides a source of water at the point where it is needed. It is owner-operated and managed
– In addition to providing an alternative source of water, rainwater harvesting systems can be used as a stormwater control system to reduce localised flooding
– It provides an essential reserve in times of emergency and/or breakdown of public water supply systems, particularly during natural disasters
– The construction of a rooftop rainwater catchment system is simple, and the community can be easily trained to build one, minimising its cost
– The technology is flexible. The systems can be built to meet almost any requirements
– It can improve the engineering of building foundations when cisterns are built as part of the substructure of the buildings, as in the case of mandatory cisterns
– The physical and chemical properties of rainwater may be superior to those of groundwater or surface waters that may have been subjected to pollution, sometimes from unknown sources
– Running costs are low, and construction, operation, and maintenance are not labour-intensive[6,7]

**Case 5.1: City of Tucson's rainwater harvesting rebate**

The City of Tucson offers a Rainwater Harvesting Rebate program to its Tucson Water customers. To be eligible for the program, applicants must have active service with Tucson Water. The program is available to single-family residential and small commercial properties. The program offers two incentive levels that applicants may apply for, and the combination of the two incentives cannot exceed $2,000 per property:

- Incentive Level 1, Simple/Passive (Rain Garden), provides a rebate of 50% of the eligible materials and labor costs up to $500. Passive earthworks practices such as basins, berms, terraces, swales, infiltration trenches, and curb cuts are covered under this incentive level. The program covers various costs, including surface, sub-surface, and conveyance features, mulch, rocks/boulders for reinforcement, soil amendments, equipment rental, permits, and labor from licensed contractors.
- Incentive Level 2, Complex/Active (Rain Tank), covers the cost of the system based on the gallon per capacity of the tank up to $2,000. The rebate is $0.25 per gallon capacity of 50–799 gallon rain tank, and $1.00 per gallon capacity of 800 gallon and larger rain tank. The rebate amount includes the tank, gutter, tank foundation, overflow, and miscellaneous materials, which make up the system.

Furthermore, applicants must attend an approved Rainwater Harvesting Incentives Program Workshop to qualify for the program.[8]

## 5.2 Rain gardens

Rain gardens are planted basins which have several functions including increasing infiltration of runoff into the ground, improving water quality by removing pollutants from the runoff, and reducing the volume of stormwater entering the stormwater management system. Rain gardens are different to regular gardens in that they usually are bowl- or saucer-shaped and are specially designed to collect runoff and hold it for one or two days as the water infiltrates into the surrounding soil. Rain gardens are often used to promote absorption and infiltration of stormwater runoff. Generally, rain gardens are most effective on a small-scale, receiving runoff from an area of no more than one-two acres: this is to avoid high volume flows which would erode plant materials. Generally, rain gardens should be between 10 and 20 percent of the square footage of the area of impervious surface that they are receiving runoff from. Rain gardens can incorporate a variety of plants including perennials, shrubs, wildflowers, and/or grasses.

Rain gardens are often placed in areas that receive runoff from a roof or paved area such as car parking lot islands, residential developments, commercial developments, and campuses. A typical rain garden includes a ponding area and inflow and outflow structures. The ponding area can be a natural or artificial ground depression, constructed by soil excavation, which, in the sloping ground, can be combined with building an earth berm at the downslope side using the excavation material. A mulch layer usually covers the ponding area's bottom before topsoil is added. If the water infiltration rate is low, a gravel layer can be constructed, or a perforated underdrain pipe installed. Inflow structures are used to convey rainwater from downspouts or adjacent impermeable areas, such as streets and footpaths, to the ponding area. The construction of overflow structures allows water to exit the rain garden when the ponding area is full, with water usually directed to the sewer network.[9,10]

Rain garden maintenance is similar to that of a typical garden, including weeding and re-establishing plants where necessary. Periodically removing sediment may be required to ensure the proper functioning of these systems. Runoff can be pre-treated via swales and/or filter strips before entering the rain garden to avoid sediment accumulation. Plants should be selected to reduce maintenance needs and to tolerate various weather conditions.[11]

In addition to managing excess runoff and improving water quality, rain gardens provide a wide range of co-benefits, for instance, they can:

- Improve the aesthetics of a property and provide a natural habitat for birds, butterflies, and beneficial insects
- Contribute to climate change adaptation by improving adaptation to more intense rainfall events that are expected to increase with climate change
- Regulate urban temperatures, reducing energy demand for cooling buildings
- Contribute to groundwater recharge
- Provide educational opportunities for schools teaching classes on sustainability and the water cycle[12,13,14,15]

**Case 5.2: The City of Melbourne's Raingarden Tree Pits**

The City of Melbourne initiated the installation of raingarden tree pits in 2006. These tree pits are well-suited for the city's central area due to their small size and flexible design options that can be tailored to suit any streetscape. In addition to cleaning stormwater runoff, the tree pits provide passive irrigation to the trees, reducing the need for manual watering. Currently, over 200 raingarden tree pits have been installed across the city, and the program is ongoing. The raingarden tree pit system works by intercepting and cleaning stormwater runoff from road surfaces that would otherwise enter the drainage system. The tree pits are placed within the kerb and contain layers of substrate that work with the tree's root system to filter pollutants, including nitrogen, phosphorus, and oils. A rock mulch layer is also included to retain moisture in the soil and prevent erosion during rain. A grate encircles the base of the tree, flush with the pavement, to protect the tree pit from damage and ensure pedestrian safety. A pipe is situated at the bottom of the well to drain the cleaned stormwater back into the drainage system. To maintain the system, street sweepers are granted access to clean litter from the inlet. Regular cleaning is also required using either manual methods or a suction hose to remove any litter or sediment build-up from the top layer of the substrate. While raingarden tree pits are designed to need minimal maintenance, periodic cleaning is still necessary. The typical annual maintenance cost of a raingarden tree pit is estimated to be between $200 and $560, which is approximately five to seven percent of the total construction cost.[16]

## 5.3 Bioswales

Bioswales are strips of vegetated areas that redirect and filter stormwater. A typical bioswale is a long, linear strip in an urban setting used to collect runoff from large impermeable surfaces such as roads and car parks. Bioswales have an inlet and outlet. When it rains, the inlet lets the water flowing down the street into the bioswale. The outlet is the kerb-cut closest to the catch basin. During heavy rain events, the bioswale may fill to its capacity, with the outlet letting excess water flow out of the bioswale so that it can flow into the catchment basin. Beneath the bioswale is layers that include sandy soils and stones that store stormwater and allow it to seep into the ground gradually. The trees and plants also absorb the stormwater and release it through evapotranspiration.

Bioswales are mainly constructed just upstream of the catch basins so that by design they can partially collect the stormwater flowing down the street and footpath before it goes into the catch basin and then into the sewer system. By partially catching stormwater in the bioswale first, the water can be used as a resource to help trees and plants grow, rather than overwhelm the sewer system. There are two types of bioswales commonly used:

– *Dry bioswales*: These provide both quantity (volume) and quality control by facilitating stormwater infiltration
– *Wet bioswales*: These use residence time and natural growth to reduce peak discharge and provide water quality treatment. A wet bioswale typically has water-tolerant vegetation permanently growing in the body of water[17,18,19]

In addition to managing excess stormwater runoff, bioswales provide a range of co-benefits including:

– Limiting the flow of water into centralised wastewater systems, therefore, reducing energy consumption and carbon emissions in wastewater treatment plants
– Increasing biodiversity in urban areas with the vegetation providing a diversity of flora that serves as a habitat for fauna
– Promoting the use of locally available construction materials
– Reducing the urban heat island effect
– Improving groundwater recharge
– Removing contaminants from stormwater
– Helping avoid polluted water from entering groundwater[20,21,22]

### 5.3.1 Maintenance of bioswales

The five main categories of care to ensure bioswales operate as they are intended are listed in Table 5.1.[23]

**Table 5.1:** Maintaining Bioswales.

| Maintenance Category | Description |
| --- | --- |
| Communicate | The difference between bioswales and regular tree pits should be communicated to residents. There should be an explanation of what they are and how they work. The more people are informed about bioswales, the better care they will take of them and the greater the acceptance of future green infrastructure systems |
| Remove rubbish | Bioswales can quickly accumulate debris and litter with rubbish easily washing in from the street, blown in from the wind, or thrown in by people. This rubbish can clog the inlets and outlets and prevent the bioswale from collecting water properly |
| Inspect the soil and plants | The soil and plants are specifically chosen to help bioswales manage stormwater. They should be inspected to make sure they are working in the way they are supposed to. Signs of soil erosion should be checked for following storms and vegetative health assessed with healthy bioswales containing healthy and dense plants |
| Weed often | Frequently weeding is essential for keeping the plants healthy. Weeds are not only unsightly, but they crowd the plants, making it hard for them to absorb water and grow strong. Weeding should be done at least once a month during the growing season |
| Sufficient water for when it is hot or dry | Even though bioswales are designed to collect rainwater, they may need additional water during hot and dry periods. Watering is necessary for the plant root systems to become established and grow |

**Case 5.3: Singapore's Sungei Serangoon Park Connector's Bioswale**

The Sungei Serangoon Park Connector in Singapore plays a crucial role in linking Punggol Park to Punggol Promenade through the Serangoon Park Connector along Sungei Pinang, providing seamless connectivity to both parks for residents. The park connector is a popular destination for North-East residents due to its aesthetically pleasing bioswale. The Sungei Serangoon Park Connector boasts a specially designed bioswale that acts as a sustainable solution for stormwater management. The bioswale is a shallow, vegetated trough or depression with a bio-retention system installed at its base, and its gently sloped sides are designed to effectively treat stormwater while preventing erosion. Additionally, the bioswale has short-term water retention capabilities, which enables efficient infiltration of surface water, making it an excellent flood control measure. With its bio-retention systems, the bioswale within the Sungei Serangoon Park Connector acts as a sustainable solution for stormwater management. The bioswale's design offers an aesthetically pleasing aspect and serves as a favorite destination for residents living in the North-East. The connector not only provides residents with seamless connectivity to both parks but also serves as an excellent example of sustainable urban planning.[24]

## 5.4 Floodwater detention and retention basins

A basin is an area that has been designed and designated for the temporary or permanent retention of floodwaters during a rain or storm event. There are two types of basins, the main difference being the presence or absence of a permanent pool of water, or pond:

- *Detention, or dry, basins*: These retain water only during storm events, releasing the water later at a controlled rate until the basin is empty. The basin remains dry between rain events
- *Retention, or wet, basins*: These retain a permanent pool of water, like a pond, irrespective of storm events and so they are wet year-round. They provide additional storage capacity above the permanent pool for the temporary storage of runoff. The depth of a wet basin is often based on water quality considerations, and so wet basins also act as water treatment devices[25]

In most cases, detention and retention basins have outlet openings that remain fixed; called static control, with the discharge rate varying according to water height but is not controlled otherwise. This static operation can be changed to a dynamic operation by controlling, in real-time, the opening of the outlet gate. The discharge rate, as well as the filling and emptying rates and volumes, can be controlled in real-time according to pre-established rules and weather and/or hydraulic conditions. Real-time control can delay the peak flow and discharge the water at times when the environment receiving the runoff has a better capacity to do so. Real-time control can also be predictive when seeking optimal solutions for discharge rates. For example, using rainfall forecasts generated by weather radars and hydrological/hydraulic models allows the system to anticipate the volume of water generated by the next rainfall event and empty the basin at the appropriate time.[26]

In addition to managing excess stormwater, detention and retention basins can improve water quality with treatment processes available including filtration, sedimentation, irradiation (UV/sun exposure), biological treatment, and plant uptake depending on the design components included in these basins.[27]

---

**Case 5.4: City of Sarnia, Ontario's Stormwater Management Ponds**
The City of Sarnia, Ontario has implemented several stormwater management ponds (SWMPs) throughout residential areas, including two in the Heritage Park subdivision, Twin Lakes, Blackwell Glen, and by the Suncor Nature Trail. SWMPs are engineered structures designed to collect and retain urban stormwater, and are often built into urban areas in North America to also retain sediments and other materials. These ponds temporarily store water and then release it at a controlled rate, providing erosion and flooding control while enhancing water quality. The increasing amount of impervious surfaces, such as roofs and roads, in urban areas reduce the time for rainfall to infiltrate before entering into the stormwater drainage system. This unchecked flow can potentially cause widespread flooding downstream. SWMPs are designed to mitigate this by containing the surge and releasing it slowly,

reducing the size and intensity of storm-induced flooding on downstream receiving waters. Furthermore, SWMPs also collect suspended sediments, which are often found in high concentrations in stormwater due to upstream construction and sand applications to roadways. By utilizing a combination of landscape and structural features, SWMPs allow sediment and contaminants to settle out of runoff before it is released into natural watercourses. The ponds also have the added benefit of holding back water to release it at a controlled rate during large storms, protecting downstream lands from erosion and flooding. Additionally, SWMPs are constructed to be an attractive feature with environmental benefits, surrounded by natural vegetation and providing habitat for birds and animals.[28]

## 5.5 Green roofs

Green roofs consist of a layer of vegetation that covers an otherwise conventional flat or moderately pitched roof. Green roofs are composed of multiple layers which can include a waterproofing roof protection layer, moisture interception layer, drainage layer, leak detection layer, engineered planting medium, and specialised plants. By appropriately selecting materials, green roofs can effectively reduce both flow peak and volume in urban drainage networks.[29] Green roofs can be used on a variety of roofs including on terraces and high-rise building roofs. Generally, any roof that has a pitch up to 16.7 percent can accommodate green roofs without special slope stabilisation provisions. There are two types of green roofs:

- *Extensive green roofs*: These are thin (usually less than six inches), lightweight systems that are mainly planted with succulents, drought-tolerant ground covering plants, and grass
- *Intensive green roofs*: These are deeper (usually greater than six inches), heavier systems that are designed to sustain complex landscapes

The typical components of green roofs include:

- *Inlet control component*: Green roofs that receive direct rainfall do not have inlet controls. For green roofs that receive runoff from roof directly connected impervious area, inlet control systems may convey and control the flow of stormwater from the contributing catchment area to the green roof
- *Storage area component*: Green roof storage areas temporarily hold stormwater before it can either be used by plants through evapotranspiration or be released downstream. Storage areas for green roofs are usually composed of:
  - *Growing medium*: This supports plant growth and provides for storage of stormwater within voids. The storage capacity is a function of medium depth, surface area, and total void space
  - *Filter or separation fabric or geotextile*: This prevents migration of soil into the underlying drainage layer of the green roof
  - *Drainage layer*: This can incorporate measures to intercept and retain percolated rainfall as it moves through the green roof storage area

- *Moisture interception layers/roof barriers*: These are impermeable liners that protect the underlying roof deck from moisture and plant root intrusion
      - *Underlying roofing system*: This typically consists of a structural deck, its supporting structures, and a traditional overlaying waterproofing system
- *Vegetation component*: Green roof plant material take up much of the water that falls on the roof during a storm event. It mitigates wind and water erosion, transpires captured moisture back into the atmosphere, and provides evaporative cooling. Plant materials also collect particulate matter and create oxygen. Some green roofs may have an irrigation system to support plant growth during dry periods
- *Outlet control component*: Outlet controls can include risers, edge drains, scuppers, gutters, or impervious liners
- *Inspection and maintenance access component*: Safe and comfortable inspection of all major components within a green roof is critical to ensuring its long-term performance. Depending on the roof height and slope, access components may consist of permanent or temporary safety monitoring systems, guardrails and safety net systems, warning line systems, and/or personal fall arrest systems. There may also be a long-term leak detection system for locating and managing leaks

In addition to managing excess stormwater runoff, green roofs provide a wide variety of co-benefits, including:
- Enhanced building aesthetics and market value
- Regulated building temperature in both the summer and winter, therefore reducing cooling and heating costs
- Reduced urban heat island effect by providing evaporative cooling
- Improved air quality by filtering particulate matter
- Extended service life of roofs by protecting the underlying roof membrane from mechanical damage, shielding it from UV radiation, and buffering temperature extremes
- Increased recreational space
- Opportunities for food production
- A wildlife habitat
- Educational resource[30,31,32,33,34,35,36]

## 5.6 Blue roofs

Blue roofs are detention systems that provide temporary storage and slow release of rainwater on a rooftop. Blue roofs can effectively control runoff from buildings with flat or mildly sloping roof surfaces. Typically, water is temporarily detained on the roof surface using rooftop check dams or rain drain restrictors. The outflow is controlled and is usually directed to the building's storm drains, scuppers or downspouts. The typical components of a blue roof include:

- *Inlet control component*: Blue roofs that only receive direct rainfall do not have inlet controls. For blue roofs that receive runoff from adjacent roof directly connected impervious area, including additional roof levels, inlet control systems convey and control the flow of stormwater from the contributing catchment area to the blue roof
- *Storage area component*: Blue roofs temporarily hold stormwater until it can either evaporate or be released downstream at a controlled rate
  - The area dedicated to storage is dependent on the chosen blue roof system type:
    - *Storage in roof drain restrictor systems*: Storage is determined by the roof slope and geometry relative to the height of both the restrictors and parapets. The bulk volume occupied by all building mechanical systems and so on need to be factored into the storage volume calculations
    - *Storage in roof check dam systems*: Storage is determined by the roof slope and associated area dedicated to ponding behind the dams. The bulk volume occupied by all building mechanical systems and so on needs to be factored into the storage volume calculations
  - In all types of blue roofs, a waterproofing membrane underlies blue roof areas with numerous types of systems existing including modified bitumen roofing, synthetic rubber membranes, thermoplastic membranes and so forth
- *Outlet control component*: Outlet controls within a blue roof system can provide a range of functions including meeting drain downtime requirements, controlling the rate of discharge and limiting water surface elevations during various storm events, and bypassing of flows from large storm events
- *Inspection and maintenance access component*: Depending on roof height and slope, blue roof inspection and maintenance access components may consist of permanent or temporary safety monitoring systems, guardrail and safety net systems, warning line systems, and/or personal fall arrest systems. There may also be a long-term leak detection system for locating and managing leaks[37]

**Case 5.5: The Hague's Green Roof Subsidy**
The Hague is offering a subsidy for green roofs with a total budget of €300,000 available for this program. Once the subsidy amount has been fully utilized, no further applications will be accepted. This subsidy program is available to homeowners, property owners (such as businesses, schools, neighborhood centers, or housing corporations), and homeowners associations (VvEs). To apply for the subsidy, homeowners and property owners must first determine whether their roof is suitable for a green roof by consulting with a certified (roofing) company. Once the green roof has been constructed, subsidy applications can be submitted. The subsidy provides €25 per square meter of green roof, up to a maximum of 50% of the construction costs, with a total subsidy of up to €20,000 per application. Several conditions must be met to qualify for the subsidy. The green roof should be constructed in accordance

with building rules, and a permit is required if a structural adaptation is needed or if the building is listed as a monument or has protected status and the green roof is visible from the street. The green roof must also have a minimum storage capacity of 18 liters per square meter, and the roof slope should not be more than 35 degrees. Furthermore, a root barrier membrane must be installed as a physical barrier to prevent root penetration. Certified companies should be consulted for the selection of the right green roofing kit.[38]

## 5.7 Permeable pavements

Permeable pavements are a stormwater practice designed to manage stormwater runoff. Permeable pavements are alternative paving surfaces that allow stormwater runoff to filter through voids in the pavement surface into an underlying stone reservoir, where the runoff is temporarily stored or infiltrated. These systems eliminate the need for construction of side drainage for collecting the stormwater. Permeable pavements also improve runoff quality as well as minimise the discharge of harmful pollutants to surface water bodies. Permeable pavements are usually limited to parking lots, basic access streets, and recreation areas which carry light vehicles or slow-moving traffic. There are four main types of categories of permeable pavement, as listed below, while Table 5.2 lists the components underneath the permeable pavement surface:

1. *Porous pavement*: This includes one or more layers of porous asphalt underlain by a choke-stone layer or treated base layer and aggregate base/sub-base reservoir. The layer depth is based on structural load, stormwater requirements, and frost depth requirements. The porous asphalt surface void space usually ranges from 18–25 percent, and surface permeability ranges from 170 to 500 inches/hour
2. *Pervious concrete*: This consists of a hydraulic cementitious binding system combined with an open-graded aggregate to produce a rigid, durable pavement. Pervious concrete pavement typically has 15–25 percent interconnected void space and a surface permeability of 300 to 2,000 inches/hour
3. *Permeable interlocking concrete pavement (PICP)*: This consists of manufactured concrete units that form permeable voids and joints when assembled into a laying pattern. The joints allow stormwater to flow into a crushed stone aggregate bedding layer and base/sub-base reservoir that support the pavers. The joints usually comprise 5–15 percent of the paver surface area and maintain surface permeability of 400–600 inches/hour
4. *Others (such as grid pavement systems)*: Grid pavements are composed of concrete or plastic open-celled paving units. The cells or openings penetrate the full thickness so they can accommodate aggregate, topsoil or grass. Surface void space ranges from 20–75 percent. Surface permeability depends on the fill material and ranges from 30 to 40 inches/hour for sand, 200 to 400 inches/hour for aggregate, and one to two inches/hour for grass fill

**Table 5.2:** Components Below Permeable Pavement Surfaces.

| Component | Description |
|---|---|
| Bedding layer | Used for pavers so they can be laid flat |
| Choker layer | A layer of small rock to prevent fine material from migrating into the reservoir layer |
| Reservoir layer | Stone to hold excess water until it infiltrates |
| Underdrain | Conveys excess water into the drainage system when the reservoir fills |
| Filter layer/geotextile | A layer of stone or permeable geotextile to separate the reservoir layer from the soil below and prevent migration of fines into the reservoir layer |
| Impermeable liner | Prevents infiltration into subgrade or adjacent roadway structural section |
| Uncompacted subgrade | Existing soil into which stormwater infiltrates |

Permeable pavements provide a range of environmental benefits, including:
– Reduced flows to storm sewer systems and streams
– Increased groundwater recharge
– Decreased and delayed peak discharge
– Reduced pollutants and improved water quality
– Reduced urban heat island effect[39,40,41,42]

**Case 5.6: City of Atlanta's Permeable Interlocking Concrete Pavement**
The Atlanta Department of Watershed Management partnered with Belgard Hardscapes in 2015–2016 to install over four miles of permeable interlocking concrete pavement (PICP) in neighbourhoods near Turner Field, home of the Atlanta Braves. The project was part of the Southeast Atlanta Green Infrastructure Initiative, aimed at managing stormwater runoff, reducing flooding, and promoting sustainability. Belgard's Aqualine L-Stone permeable paver was chosen for the project due to its durability and ability to withstand vehicular traffic in high-traffic areas. The pavement also delivered results in both cost-effectiveness and long-term durability in harsh climates, particularly those with extreme freeze/thaw cycles. The installation of the permeable pavers involved removing the asphalt, excavating the street, filling it with stone and a terraced dam system, and installing the pavers on top of the stone. The unique "L-shape" of the stone allows for minimal waste during installation and provides optimal interlock for vehicular traffic. As of June 2016, the new PICP system was providing over two million gallons of storage capacity, giving residents extra flood protection during heavy rain events.[43]

## 5.8 Green streets

Green streets use green infrastructure practices installed within the public right-of-way to manage stormwater while preserving the primary function of a street as a conduit for vehicles, pedestrians, cyclists, and transit riders. Green streets also reduce contaminants from entering local waterways, improving water quality. The various types of green infrastructure commonly used in transitways, boulevards, main neighbourhood streets, commercial and residential shared streets, green alleys, industrial streets, and so forth, are listed below. At the same time, Table 5.3 provides a summary of the multiple benefits green streets provide.

**Table 5.3:** Benefits of Green Streets.

| Street User | Green Street Benefit |
|---|---|
| People walking | – Make the walking environment more inviting and pleasant by reducing the temperature, attenuating noise, and improving air quality<br>– Calm traffic and improve safety conditions<br>– High-quality public gathering spaces with natural features improve mental health and create opportunities for community development and social cohesion |
| People using transit | – Green infrastructure can be integrated into transit facilities, including traffic islands to improve natural drainage near transit stops<br>– Transit shelters and facility roofs can incorporate green infrastructure<br>– Green infrastructure can be incorporated alongside cycleways to improve drainage and increase cycling comfort and access during and after storms<br>– Permeable pavement can be implemented on cycle lanes and raised cycle tracks to reduce the time required for the pavement to dry<br>– Planters or vegetation may be incorporated into protected cycleway buffer elements to increase ride comfort and reduce stress |
| People driving motor vehicles | – Green infrastructure can capture runoff and reduce flooding and ponding, promoting safer driving conditions<br>– Green infrastructure can be implemented with geometric changes that reduce vehicle speed and improve visibility |
| People conducting business | – Success and viability of commercial districts and neighbourhood shops depends on the ability of people to access and use streets comfortably<br>– Economic performance is tied to the comfort and attractiveness of streets, with environments with green infrastructure performing better<br>– Green infrastructure can increase property value |

### 5.8.1 Stormwater planters

A stormwater planter is a specialised, landscaped planter installed in the footpath area and are designed to manage stormwater runoff. Runoff is routed to the planter by setting the top of the planting media in the planter lower than the street's gutter elevation and connecting the planter to one or more inlets, allowing stormwater runoff from the street to flow into the planter. Runoff from the adjacent footpath can flow directly into the stormwater planter from the surface. Plantings are incorporated within the system to provide uptake of water and pollutants.

### 5.8.2  Stormwater bump-outs

A stormwater bump-out is a landscape kerb extension that extends the existing kerb line into the cartway. It is designed to manage stormwater runoff by setting the top of the planting media in the bump-out lower than the street's gutter elevation and connecting the bump-out to one or more inlets, allowing stormwater runoff from the street to flow into the bump-outs. Runoff from the adjacent footpath can flow directly into the stormwater bump-out from the surface. Stormwater bump-outs capture, slow, and infiltrate stormwater within a planted area or subsurface stone bed. Plantings take up some of the stormwater through their root systems, and the remaining stormwater is temporarily stored within the kerb extension until it either infiltrates or drains back to the sewer.

### 5.8.3  Stormwater tree

A stormwater tree is a street tree planted in a specialised tree pit installed in the footpath area. It is designed to manage stormwater runoff by placing the top of the planting media in the tree pit lower than the street's gutter elevation and connecting the tree pit to an inlet, allowing stormwater runoff from the street to flow into the tree pit. Runoff from the adjacent footpath can flow directly into the tree pit from the footpath surface. If the stormwater tree pit reaches capacity, runoff can bypass the inlet and enter other downstream green infrastructure or a storm drain.

### 5.8.4  Stormwater tree trench

A stormwater tree trench is a subsurface trench installed in the footpath area that includes a series of trees along a section or the total length of the subsurface trench. It manages stormwater runoff by connecting the subsurface trench to one or more inlets, allowing runoff from the street and footpath to flow into the subsurface trench. The runoff is stored in the empty spaces between the stones or other storage media in the trench, watering the trees and slowly infiltrating through the trench bottom.

### 5.8.5  Green car parking lots

Green car parking lots incorporate a variety of green infrastructure design elements, including trees, dispersion areas, bioinfiltration, and permeable pavement. These strategies use natural processes to reduce the volume of runoff, peak flow, and pollutants. In particular:

- *Trees*: Trees intercept water on leaves, slowly delivering it to mulch and soils, absorbing it through root systems, and transpiring it as water vapour directly back to the atmosphere
- *Dispersion areas*: Dispersion areas disconnect impervious areas from directly running to the storm drainage system. Dispersion areas use the natural functions of plants, mulch, and soils to slow stormwater runoff and remove pollutants. This strategy uses storage, sediment capture, and biological processes to clean the water
- *Bioinfiltration*: Bioinfiltration facilities are vegetated surface water systems that filter water through vegetation and soil or bioinfiltration soil media before discharge to the storm drain system. They also use shallow depressions to provide storage and evapotranspiration[44,45,46,47,48,49,50,51,52]

**Case 5.7: City of Toronto's Green Streets**
The City of Toronto has completed various green streets projects with more in the planning stage. The projects are carried out directly by the city or through development-led initiatives. The city evaluates and prioritizes streets and locations suitable for green streets using a technical screening process, considering factors such as soil quality and infiltration rates, tree canopy, stormwater management, social wellness and equity, and impact on mature trees. The implementation of green streets is often combined with planned construction to reduce construction costs and minimize disruption to the community. As of May 2022, the Toronto Green Standard Version 4 mandates the inclusion of green infrastructure in all newly constructed streets. Green streets systems are maintained by the city, designated contractors, property owners, or through partnerships with social enterprises. In 2021, Green Streets launched the GreenForceTO pilot project to create local green jobs in landscaping and property maintenance. The initiative aims to build employment skills and develop career pathways for individuals experiencing barriers to employment. The city partnered with two local employment social enterprises, RAINscapeTO and Building Up, to maintain bio-swales, pollinator gardens, and other green spaces that improve neighbourhood climate resilience and biodiversity.[53]

## 5.9 Multifunctional spaces

Multifunctional spaces, including streets, parking spaces, green spaces, sports grounds, and playgrounds can be used for short-term retention and/or transportation of runoff peaks during extreme precipitation events. These multifunctional spaces provide numerous benefits, including:
- Reductions in impervious areas
- Infiltration of runoff from paved areas and rooftops
- Public education opportunities
- Provision of shade when trees are used
- Improved habitat for wildlife
- Creation of a more welcoming environment
- Creation of park-like areas[54,55]

**Case 5.8: Hamburg's Stadium Storing Rainwater Underground**

The City of Hamburg, Hamburg Wasser, and the State Ministry for Urban Development and Environment are working on the Rain InfraStructure Adaptation project (Project RISA), which aims to provide stormwater management solutions to prevent flooding and water pollution from sewer overflow and street runoff. Project RISA encourages the use of multifunctional spaces including streets, parking spaces, green spaces, sports grounds, and playgrounds for short-term retention and/or transportation of runoff peaks during extreme precipitation events. The use of public spaces is limited to once-every-five-year events and streets once-every-ten-year events. The design of multifunctional spaces considers the extent to which each area is potentially at risk of flooding and the intensity of use. As part of Project RISA, Hamburg Wasser is constructing an underground storage and infiltration system under the Hein Klink Stadium in Billstedt to prevent localised flooding. The excess water from heavy rain will flow through a newly laid channel over a settling shaft into trenches under the stadium. In the first step of construction, trenches were built under the sports field to absorb excess water and gradually release it to the ground. In the next construction step, excess water will be led from the street to the sports field via a siel and a settling shaft. The system will be able to absorb more than 500,000 litres of water during a heavy rainfall event. Once completed, new surfaces will be applied to the sports field.[56]

# Notes

1  R.C. Brears, *Blue and Green Cities: The Role of Blue-Green Infrastructure in Managing Urban Water Resources* (Palgrave Macmillan UK, 2018).

2  Violet Kisakye and Bart Van der Bruggen, "Effects of Climate Change on Water Savings and Water Security from Rainwater Harvesting Systems," Resources, *Conservation and Recycling* 138 (2018).

3  Mohammad A. Alim et al., "Suitability of Roof Harvested Rainwater for Potential Potable Water Production: A Scoping Review," *Journal of Cleaner Production* 248 (2020).

4  N. İpek Şahin and Gülten Manioğlu, "Water Conservation through Rainwater Harvesting Using Different Building Forms in Different Climatic Regions," *Sustainable Cities and Society* 44 (2019).

5  B. Helmreich and H. Horn, "Opportunities in Rainwater Harvesting," *Desalination* 248, no. 1 (2009).

6  Jennifer Steffen et al., "Water Supply and Stormwater Management Benefits of Residential Rainwater Harvesting in U.S. Cities," *JAWRA Journal of the American Water Resources Association* 49, no. 4 (2013).

7  UNEP, "Source Book of Alternative Technologies for Freshwater Augmentation in Latin America and the Caribbean," https://www.oas.org/dsd/publications/Unit/oea59e/ch20.htm#2.1%20desalination%20by%20reverse%20osmosis.

8  City of Tucson, "Conservation," https://www.tucsonaz.gov/Departments/Water/Conservation.

9  Aikaterini Basdeki, Lysandros Katsifarakis, and Konstantinos L. Katsifarakis, "Rain Gardens as Integral Parts of Urban Sewage Systems-a Case Study in Thessaloniki, Greece," *Procedia Engineering* 162 (2016).

10  Laurène Autixier et al., "Evaluating Rain Gardens as a Method to Reduce the Impact of Sewer Overflows in Sources of Drinking Water," *Science of The Total Environment* 499 (2014).

11  City of Chicago, "City of Chicago Bioinfiltration Rain Gardens," https://www.chicago.gov/city/en/depts/water/supp_info/conservation/green_design/bioinfiltration_raingardens.html.

12  Naturally Resilient Communities, "Rain Gardens," http://nrcsolutions.org/rain-gardens/.

13  Natural Water Retention Measures, "Rain Gardens," (2015), http://nwrm.eu/measure/rain-gardens.

**14** Siwiec Ewelina, Erlandsen Anne Maren, and Vennemo Haakon, "City Greening by Rain Gardens – Costs and Benefits," *Environmental Protection and Natural Resources; The Journal of Institute of Environmental Protection-National Research Institute.* 29, no. 1 (2018).

**15** Sarah P. Church, "Exploring Green Streets and Rain Gardens as Instances of Small Scale Nature and Environmental Learning Tools," *Landscape and Urban Planning* 134 (2015).

**16** City of Melbourne, "Raingarden Tree Pit Program," https://urbanwater.melbourne.vic.gov.au/proj ects/raingardens/little-collins-street-tree-pits/.

**17** University of Florida, "Bioswales/Vegetated Swales," (2008), http://buildgreen.ufl.edu/Fact_sheet_bi oswales_Vegetated_Swales.pdf.

**18** Brears, *Blue and Green Cities: The Role of Blue-Green Infrastructure in Managing Urban Water Resources.*

**19** *Nature-Based Solutions to 21st Century Challenges* (Oxfordshire, UK: Routledge, 2020).

**20** CTCN, "Bioswales," (2017), https://www.ctc-n.org/resources/bioswales.

**21** A. Rebecca Purvis et al., "Evaluating the Water Quality Benefits of a Bioswale in Brunswick County, North Carolina (Nc), USA," *Water* 10, no. 2 (2018).

**22** Brian S. Anderson et al., "Bioswales Reduce Contaminants Associated with Toxicity in Urban Storm Water," *Environmental Toxicology and Chemistry* 35, no. 12 (2016).

**23** G. Everett et al., "Delivering Green Streets: An Exploration of Changing Perceptions and Behaviours over Time around Bioswales in Portland, Oregon," *Journal of Flood Risk Management* 11, no. S2 (2018).

**24** NParks, "Sungei Serangoon Pc," https://www.nparks.gov.sg/gardens-parks-and-nature/park-connec tor-network/sungei-serangoon-pc.

**25** Naturally Resilient Communities, "Floodwater Detention and Retention Basins," (2017), http://nrcso lutions.org/wp-content/uploads/2017/03/NRC_Solutions_Retention_Basins.pdf.

**26** Karine Bilodeau, Geneviève Pelletier, and Sophie Duchesne, "Real-Time Control of Stormwater Detention Basins as an Adaptation Measure in Mid-Size Cities," *Urban Water Journal* 15, no. 9 (2018).

**27** David B. E. Pezzaniti, Simon Gche M. E. PhD Beecham, and Jaya M. E. PhD Kandasamy, "Stormwater Detention Basin for Improving Road-Runoff Quality," *Proceedings of the Institution of Civil Engineers* 165, no. 9 (2012).

**28** Ontario City of Sarnia, "Stormwater Management Ponds," https://www.sarnia.ca/stormwater-man agement-ponds/#:~:text=A%20stormwater%20management%20pond%20%28SWMP%29%20is%20an% 20artificial,constructed%20to%20gather%20rainfall%20and%20surface%20water%20runoff.

**29** Giulia Ercolani et al., "Evaluating Performances of Green Roofs for Stormwater Runoff Mitigation in a High Flood Risk Urban Catchment," *Journal of Hydrology* 566 (2018).

**30** Muhammad Shafique, Reeho Kim, and Muhammad Rafiq, "Green Roof Benefits, Opportunities and Challenges – a Review," *Renewable and Sustainable Energy Reviews* 90 (2018).

**31** Philadelphia Water Department, "Stormwater Management Practice Guidance," (2018), https:// www.pwdplanreview.org/manual-info/guidance-manual.

**32** Lotte Fjendbo Møller Francis and Marina Bergen Jensen, "Benefits of Green Roofs: A Systematic Review of the Evidence for Three Ecosystem Services," *Urban Forestry & Urban Greening* 28 (2017).

**33** Shafique, Kim, and Rafiq, "Green Roof Benefits, Opportunities and Challenges – a Review."

**34** Ahmet B. Besir and Erdem Cuce, "Green Roofs and Facades: A Comprehensive Review," ibid.82.

**35** Virginia Stovin, "The Potential of Green Roofs to Manage Urban Stormwater," *Water and Environment Journal* 24, no. 3 (2010).

**36** T. Susca, S. R. Gaffin, and G. R. Dell'Osso, "Positive Effects of Vegetation: Urban Heat Island and Green Roofs," *Environmental Pollution* 159, no. 8 (2011).

**37** Philadelphia Water Department, "Stormwater Management Practice Guidance".

**38** The Hague, "Apply for a Climate Adaptation Subsidy (for Rainwater Harvesting)," https://www.den haag.nl/en/subsidies/apply-for-a-climate-adaptation-subsidy-for-rainwater-harvesting-.htm.

**39** Shadi Saadeh et al., "Application of Fully Permeable Pavements as a Sustainable Approach for Mitigation of Stormwater Runoff," *International Journal of Transportation Science and Technology* (2019).

**40** Masoud Kayhanian et al., "Application of Permeable Pavements in Highways for Stormwater Runoff Management and Pollution Prevention: California Research Experiences," ibid.

**41** Transportation Research Board, Engineering National Academies of Sciences, and Medicine, *Guidance for Usage of Permeable Pavement at Airports*, ed. James Bruinsma, et al. (Washington, DC: The National Academies Press, 2017).

**42** San Diego County, "Green Parking Lots Guidelines: A Guide to Green Parking Lots Implementation in the County of San Diego" (2019), https://www.sandiegocounty.gov/content/dam/sdc/dpw/WATER SHED_PROTECTION_PROGRAM/watershedpdf/Dev_Sup/GPL_Guidelines_2019.pdf.

**43** Belgard, "Case Study: City of Atlanta Permeable Paver Retrofit," https://www.belgardcommercial. com/case_studies/case_study_city_of_atlanta_permeable_paver_retrofit#.

**44** Philadelphia Water Department, "City of Philadelphia Green Streets Design Manual," (2014), http:// www.phillywatersheds.org/img/GSDM/GSDM_FINAL_20140211.pdf.

**45** Adam Berland et al., "The Role of Trees in Urban Stormwater Management," *Landscape and urban planning* 162 (2017).

**46** National Association of City Transportation Officials, "Urban Street Stormwater Guide," (2017), https://nacto.org/publication/urban-street-stormwater-guide/.

**47** Ibid.

**48** State of New Jersey, "Complete and Green Streets for All. Model Complete Streets Policy & Guide Making New Jersey's Communities Healthy, Equitable, Green & Prosperous," (2019), https://www.state. nj.us/transportation/eng/completestreets/pdf/CS_Model_Policy_2019.pdf.

**49** Joowon Im, "Green Streets to Serve Urban Sustainability: Benefits and Typology," *Sustainability* 11, no. 22 (2019).

**50** David Elkin, "Portland's Green Streets: Lessons Learned Retrofitting Our Urban Watersheds," in *Low Impact Development for Urban Ecosystem and Habitat Protection* (2008).

**51** Guillem Vich, Oriol Marquet, and Carme Miralles-Guasch, "Green Streetscape and Walking: Exploring Active Mobility Patterns in Dense and Compact Cities," *Journal of Transport & Health* 12 (2019).

**52** San Diego County, "Green Parking Lots Guidelines: A Guide to Green Parking Lots Implementation in the County of San Diego ".

**53** City of Toronto, "Green Streets Projects," https://www.toronto.ca/services-payments/streets-park ing-transportation/enhancing-our-streets-and-public-realm/green-streets/green-streets-projects/.

**54** Brears, *Blue and Green Cities: The Role of Blue-Green Infrastructure in Managing Urban Water Resources*.

**55** *Nature-Based Solutions to 21st Century Challenges*.

**56** Robert C. Brears, "Hamburg's Stadium Storing Rainwater Underground," Mark and Focus, https:// medium.com/mark-and-focus/hamburgs-stadium-storing-rainwater-underground-8b44e3941e15.

# References

Alim, Mohammad A., Ataur Rahman, Zhong Tao, Bijan Samali, Muhammad M. Khan, and Shafiq Shirin. "Suitability of Roof Harvested Rainwater for Potential Potable Water Production: A Scoping Review". *Journal of Cleaner Production* 248 (2020/03/01/ 2020): 119226.

Anderson, Brian S., Bryn M. Phillips, Jennifer P. Voorhees, Katie Siegler, and Ronald Tjeerdema. "Bioswales Reduce Contaminants Associated with Toxicity in Urban Storm Water". *Environmental Toxicology and Chemistry* 35, no. 12 (2016/12/01 2016): 3124–34.

Autixier, Laurène, Alain Mailhot, Samuel Bolduc, Anne-Sophie Madoux-Humery, Martine Galarneau, Michèle Prévost, and Sarah Dorner. "Evaluating Rain Gardens as a Method to Reduce the Impact of Sewer Overflows in Sources of Drinking Water". *Science of The Total Environment* 499 (2014/11/15/ 2014): 238–47.

Basdeki, Aikaterini, Lysandros Katsifarakis, and Konstantinos L. Katsifarakis. "Rain Gardens as Integral Parts of Urban Sewage Systems-a Case Study in Thessaloniki, Greece". *Procedia Engineering* 162 (2016/01/01/ 2016): 426–32.

Basler & Hofmann. "Flood Water Retention for the City of Winterthur". https://www.baslerhofmann.ch/en/ projects/en-projekte-detailansicht/projekt/hochwasserrueckhalt-fuer-die-stadt-winterthur.html.

Belgrade. "Belgard Partners with the City of Atlanta for the Largest Permeable Pavement Project". https://www.belgardcommercial.com/resources/news_and_articles/belgard_partners_with_the_city_ of_atlanta.

Berland, Adam, Sheri A. Shiflett, William D. Shuster, Ahjond S. Garmestani, Haynes C. Goddard, Dustin L. Herrmann, and Matthew E. Hopton. "The Role of Trees in Urban Stormwater Management". [In eng]. *Landscape and urban planning* 162 (2017): 167–77.

Besir, Ahmet B., and Erdem Cuce. "Green Roofs and Facades: A Comprehensive Review". *Renewable and Sustainable Energy Reviews* 82 (2018/02/01/ 2018): 915–39.

Bilodeau, Karine, Geneviève Pelletier, and Sophie Duchesne. "Real-Time Control of Stormwater Detention Basins as an Adaptation Measure in Mid-Size Cities". *Urban Water Journal* 15, no. 9 (2018/10/21 2018): 858–67.

Board, Transportation Research, Engineering National Academies of Sciences, and Medicine. *Guidance for Usage of Permeable Pavement at Airports*. [in English] Edited by James Bruinsma, Kelly Smith, David Peshkin, Lauren Ballou, Bethany Eisenberg, Carol Lurie, Mark Costa, *et al*. Washington, DC: The National Academies Press, 2017. doi:10.17226/24852.

Brears, R.C. Blue and Green Cities: The Role of Blue-Green Infrastructure in Managing Urban Water Resources. Palgrave Macmillan UK, 2018.

_____. *Nature-Based Solutions to 21st Century Challenges*. Oxfordshire, UK: Routledge, 2020.

Church, Sarah P. "Exploring Green Streets and Rain Gardens as Instances of Small Scale Nature and Environmental Learning Tools". *Landscape and Urban Planning* 134 (2// 2015): 229–40.

City of Chicago. "City of Chicago Bioinfiltration Rain Gardens". https://www.chicago.gov/city/en/depts/ water/supp_info/conservation/green_design/bioinfiltration_raingardens.html.

City of Toronto. "Earl Bales Stormwater Management Pond". https://www.toronto.ca/services-payments/ water-environment/managing-rain-melted-snow/what-the-city-is-doing-stormwater-management- projects/other-stormwater-management-projects/stormwater-ponds/earl-bales-stormwater- management-pond/.

_____. "Green Street Technical Guidelines" (2017). https://www.toronto.ca/services-payments/streets- parking-transportation/enhancing-our-streets-and-public-realm/green-streets/.

City of Tucson. "Rainwater Harvesting Rebate". https://www.tucsonaz.gov/water/rainwater-harvesting- rebate.

CTCN. "Bioswales". (2017). https://www.ctc-n.org/resources/bioswales.

Elkin, David. "Portland's Green Streets: Lessons Learned Retrofitting Our Urban Watersheds". In *Low Impact Development for Urban Ecosystem and Habitat Protection*, 1–9, 2008.

Ercolani, Giulia, Enrico Antonio Chiaradia, Claudio Gandolfi, Fabio Castelli, and Daniele Masseroni. "Evaluating Performances of Green Roofs for Stormwater Runoff Mitigation in a High Flood Risk Urban Catchment". *Journal of Hydrology* 566 (2018/11/01/ 2018): 830–45.

Everett, G., J. E. Lamond, A. T. Morzillo, A. M. Matsler, and F. K. S. Chan. "Delivering Green Streets: An Exploration of Changing Perceptions and Behaviours over Time around Bioswales in Portland, Oregon". *Journal of Flood Risk Management* 11, no. S2 (2018/02/01 2018): S973–S85.

Ewelina, Siwiec, Erlandsen Anne Maren, and Vennemo Haakon. "City Greening by Rain Gardens – Costs and Benefits". [In English]. *Environmental Protection and Natural Resources; The Journal of Institute of Environmental Protection-National Research Institute*. 29, no. 1 (2018): 1–5.

Francis, Lotte Fjendbo Møller, and Marina Bergen Jensen. "Benefits of Green Roofs: A Systematic Review of the Evidence for Three Ecosystem Services". *Urban Forestry & Urban Greening* 28 (2017/12/01/ 2017): 167–76.

Helmreich, B., and H. Horn. "Opportunities in Rainwater Harvesting". *Desalination* 248, no. 1 (2009/11/15/ 2009): 118–24.

Im, Joowon. "Green Streets to Serve Urban Sustainability: Benefits and Typology". *Sustainability* 11, no. 22 (2019).

Kayhanian, Masoud, Hui Li, John T. Harvey, and Xiao Liang. "Application of Permeable Pavements in Highways for Stormwater Runoff Management and Pollution Prevention: California Research Experiences". *International Journal of Transportation Science and Technology* (2019/02/02/ 2019).

Kisakye, Violet, and Bart Van der Bruggen. "Effects of Climate Change on Water Savings and Water Security from Rainwater Harvesting Systems". *Resources, Conservation and Recycling* 138 (2018/11/01/ 2018): 49–63.

National Association of City Transportation Officials. "Urban Street Stormwater Guide". (2017). Natural Water Retention Measures. "Rain Gardens". (2015). http://nwrm.eu/measure/rain-gardens.

Naturally Resilient Communities. "Floodwater Detention and Retention Basins". (2017). http://nrcsolutions. org/wp-content/uploads/2017/03/NRC_Solutions_Retention_Basins.pdf.

————. "Rain Gardens". http://nrcsolutions.org/rain-gardens/.

Pezzaniti, David B. E., Simon Gche M. E. PhD Beecham, and Jaya M. E. PhD Kandasamy. "Stormwater Detention Basin for Improving Road-Runoff Quality". [In English]. *Proceedings of the Institution of Civil Engineers* 165, no. 9 (Oct 2012 2012-09-27 2012): 461–71.

Philadelphia Water Department. "City of Philadelphia Green Streets Design Manual". (2014). http://www. phillywatersheds.org/img/GSDM/GSDM_FINAL_20140211.pdf.

————. "Stormwater Management Practice Guidance". (2018). https://www.pwdplanreview.org/manual-info/guidance-manual.

PUB. "Kallang River @ Bishan-Ang Mo Kio Park". https://www.pub.gov.sg/abcwaters/explore/bishanangmokiopark.

Purvis, A. Rebecca, J. Ryan Winston, F. William Hunt, Brian Lipscomb, Karthik Narayanaswamy, Andrew McDaniel, S. Matthew Lauffer, and Susan Libes. "Evaluating the Water Quality Benefits of a Bioswale in Brunswick County, North Carolina (Nc), USA". *Water* 10, no. 2 (2018).

Saadeh, Shadi, Avinash Ralla, Yazan Al-Zubi, Rongzong Wu, and John Harvey. "Application of Fully Permeable Pavements as a Sustainable Approach for Mitigation of Stormwater Runoff". *International Journal of Transportation Science and Technology* (2019/02/13/ 2019).

Şahin, N. İpek, and Gülten Manioğlu. "Water Conservation through Rainwater Harvesting Using Different Building Forms in Different Climatic Regions". *Sustainable Cities and Society* 44 (2019/01/01/ 2019): 367–77.

San Diego County. "Green Parking Lots Guidelines: A Guide to Green Parking Lots Implementation in the County of San Diego" (2019). https://www.sandiegocounty.gov/content/dam/sdc/dpw/WATERSHED_PROTECTION_PROGRAM/watershedpdf/Dev_Sup/GPL_Guidelines_2019.pdf.

SFPUC. "San Francisco Rain Guardians". https://sfwater.org/index.aspx?page=1190.

Shafique, Muhammad, Reeho Kim, and Muhammad Rafiq. "Green Roof Benefits, Opportunities and Challenges – a Review". *Renewable and Sustainable Energy Reviews* 90 (2018/07/01/ 2018): 757–73.

State of New Jersey. "Complete and Green Streets for All. Model Complete Streets Policy & Guide Making New Jersey's Communities Healthy, Equitable, Green & Prosperous". (2019). https://www.state.nj.us/transportation/eng/completestreets/pdf/CS_Model_Policy_2019.pdf.

Steffen, Jennifer, Mark Jensen, Christine A. Pomeroy, and Steven J. Burian. "Water Supply and Stormwater Management Benefits of Residential Rainwater Harvesting in U.S. Cities". *JAWRA Journal of the American Water Resources Association* 49, no. 4 (2013): 810–24.

Stovin, Virginia. "The Potential of Green Roofs to Manage Urban Stormwater". *Water and Environment Journal* 24, no. 3 (2010): 192–99.

Susca, T., S. R. Gaffin, and G. R. Dell'Osso. "Positive Effects of Vegetation: Urban Heat Island and Green Roofs". *Environmental Pollution* 159, no. 8 (2011/08/01/ 2011): 2119–26.

UNEP. "Source Book of Alternative Technologies for Freshwater Augmentation in Latin America and the Caribbean". https://www.oas.org/dsd/publications/Unit/oea59e/ch20.htm#2.1%20desalination%20by %20reverse%20osmosis.

University of Florida. "Bioswales/Vegetated Swales". (2008). http://buildgreen.ufl.edu/Fact_sheet_bio swales_Vegetated_Swales.pdf.

Urban Innovation Actions. "Resilio – Resilience Network of Smart Innovative Climate-Adapative Rooftops". https://www.uia-initiative.eu/en/uia-cities/amsterdam.

Vich, Guillem, Oriol Marquet, and Carme Miralles-Guasch. "Green Streetscape and Walking: Exploring Active Mobility Patterns in Dense and Compact Cities". *Journal of Transport & Health* 12 (2019/03/01/ 2019): 50–59.

Belgard. "Case Study: City of Atlanta Permeable Paver Retrofit." https://www.belgardcommercial.com/ case_studies/case_study_city_of_atlanta_permeable_paver_retrofit#.

City of Melbourne. "Raingarden Tree Pit Program." https://urbanwater.melbourne.vic.gov.au/projects/rain gardens/little-collins-street-tree-pits/.

City of Sarnia, Ontario. "Stormwater Management Ponds." https://www.sarnia.ca/stormwater- management-ponds/#:~:text=A%20stormwater%20management%20pond%20%28SWMP%29%20is% 20an%20artificial,constructed%20to%20gather%20rainfall%20and%20surface%20water%20runoff.

——. "Green Streets Projects." https://www.toronto.ca/services-payments/streets-parking-transportation/ enhancing-our-streets-and-public-realm/green-streets/green-streets-projects/.

City of Tucson. "Conservation." https://www.tucsonaz.gov/Departments/Water/Conservation.

NParks. "Sungei Serangoon Pc." https://www.nparks.gov.sg/gardens-parks-and-nature/park-connector- network/sungei-serangoon-pc.

Robert C. Brears. "Hamburg's Stadium Storing Rainwater Underground." Mark and Focus, https://medium. com/mark-and-focus/hamburgs-stadium-storing-rainwater-underground-8b44e3941e15.

The Hague. "Apply for a Climate Adaptation Subsidy (for Rainwater Harvesting)." https://www.denhaag.nl/ en/subsidies/apply-for-a-climate-adaptation-subsidy-for-rainwater-harvesting-.htm.

Robert C. Brears. "Hamburg's Stadium Storing Rainwater Underground." Mark and Focus, https://medium. com/mark-and-focus/hamburgs-stadium-storing-rainwater-underground-8b44e3941e15.

The Hague. "Apply for a Climate Adaptation Subsidy (for Rainwater Harvesting)." https://www.denhaag.nl/ en/subsidies/apply-for-a-climate-adaptation-subsidy-for-rainwater-harvesting-.htm.

# Chapter 6
# Protecting and restoring water quality in river basins

**Abstract:** River basins are geographic features that include all surface and groundwater, soils, vegetation, animals, and human activities and do not reflect local political boundaries. In most cases, river basins cross political and administrative boundaries and so by cooperating, communities within river basins can plan for the future of the river basin. River basin management focuses on the relationship between land use and land cover, the movement and storage of water, and water quality. This chapter will discuss how river basin planning can protect and restore water quality before discussing how permit systems, best management practices, and source water protection can protect and restore water quality.

**Keywords:** River Basin Management, Tradable Permits, Water Quality, Best Management Practices

## Introduction

River basins are geographic features that include all surface and groundwater, soils, vegetation, animals, and human activities and do not reflect local political boundaries. In most cases, river basins cross political and administrative boundaries and so by cooperating, communities within river basins can plan for the future of the river basin. River basin management focuses on the relationship between land use and land cover, the movement and storage of water, and water quality. This chapter will discuss how river basin planning can protect and restore water quality before discussing how permit systems, best management practices, and source water protection can protect and restore water quality.

## 6.1  River basin planning to protect and restore water quality

A successful river basin plan to protect and restore water quality should clearly identify why the river basin plan is needed, where the existing problems, threats, and opportunities are located, what actions and projects are recommended to address the problems and threats and to take advantage of the opportunities, when the recommendations will be advanced, who will take the lead in making it happen, and how much will it cost to implement the plan. The river basin plan will have a variety of goals and strategies that it seeks to maintain or achieve. A successful river basin plan

https://doi.org/10.1515/9783111028101-006

is informed by available data and driven by consensus. The plan characterises the physical aspects of the river basin and assesses municipal controls to ensure future water quality. The plan will also recommend corrective and preventive actions to protect and restore water quality as well as other ecosystem services.

### 6.1.1 Developing and implementing a successful river basin management plan

There are a variety of recommended activities that should take place to ensure the success of a river basin management plan as follows.

#### 6.1.1.1 Planning for community involvement

River basin planning is only successful when the people that live and work in the river basin realise that they are a crucial part of their river basin. They recognise that their actions have an impact on the health of the river basin and participate in protecting and restoring their river basin for the benefit of the entire community and future generations. Community involvement is required in every stage of protecting and restoring the river basin. Community participation can take many forms, but it is generally designed to:
–   Foster an appreciation of the river basin
–   Introduce local leaders and community residents to the river basin planning process
–   Generate a community consensus about the vision for the future of the river basin
–   Develop a strategy to address the most critical river basin issues

#### 6.1.1.2 Identifying the key stakeholders

A stakeholder is a person or group who has something to gain or lose based on the outcomes of the river basin plan. It is essential to involve those individuals and groups, which include elected officials, business and civic leaders, neighbourhood and environmental groups, and educational institutions, who have a direct stake in the future of the river basin. The setting and prioritising of goals are where stakeholders become critical players in identifying strategies and designing the actual river basin management plan. A strategy needs to be formed that effectively involves these stakeholders. The strategy needs to identify how to approach each stakeholder and what they can contribute, for instance:
–   Do they need to be informed?
–   Are the stakeholders required for funding?
–   Are they to participate on a committee?

Throughout the river basin plan's implementation, stakeholders and other community members may be involved in a variety of ways, for example, creating a river basin advisory committee, creating specific issue-orientated subcommittees, establishing partnerships, encouraging partnerships, encouraging participating in visioning and planning workshops, or participating in volunteer work parties. At the same time, the entire community needs to be informed on what is going on with regular progress reports provided and how they can participate.

### 6.1.1.3 Organising a river basin advisory committee

In the early stages of the planning process, most communities establish an advisory committee that helps focus efforts, streamlines the planning process, and see the plan through to implementation. Responsibilities of advisory committees usually include:
- Managing the river basin planning process or advising staff on managing the process
- Providing input on river basin issues and conditions
- Holding regular meetings related to planning and project implementation
- Informing the community about the planning process and ways they can be involved
- Organising and participating in focus groups, workshops, and public hearings
- Keeping elected officials and municipal officials informed about the planning process

Members of the advisory committee should include representatives from groups including local governments, such as elected officials, staff, and members of planning, zoning and other boards of all municipalities in the river basin, neighbourhood and community organisations, local and regional non-profit organisations, property owners, representatives from the business community, water suppliers, and the academic community.

### 6.1.1.4 Establishing partnerships

In addition to advisory committees bringing together all the stakeholders in a river basin, partnerships need to be formed with these stakeholders. Partnerships are required when pursuing goals that affect other people and organisations, more resources, whether financial, political, or human, are required to accomplish goals, and a strong coalition is required to show that interests of various stakeholders are in agreement. Success in river basin planning requires partnerships that contain the right blend of stakeholders. They will come from the private sector, all levels of government, and from the community to form a partnership with the common goal of achieving a shared vision. There are a variety of critical potential partners, including local government, adjacent municipalities, regional planning or resource conservation organisations, state government partners, academic institutions, representatives of businesses and industries in the surrounding area, property owners in the surrounding area, residents in the surrounding area, community and neighbourhood groups, and non-profit organisation in the community and river basin.

### 6.1.1.5 Charting the course

It is essential to chart the course for the protection and restoration of the river basin by developing a step-by-step strategy to guide the completion and implementation of the river basin plan. To chart the course, the following needs to be specified:

– The tasks to be performed
– The technique to be used
– The roles of the people involved and their areas of responsibility
– The time frame for action

### 6.1.1.6 Implementing goals and strategies and monitoring

To fit their river basin's needs, stakeholders and decision-makers may customise the tools that exist for putting river basin management plans into practice. Several of those tools are permits, best management practices (BMPs), and source water protection. Each river basin management plan will have site-specific needs requiring different combinations of these tools. Measuring progress towards achieving river basin management plans and water quality goals can be done through increased and more efficient monitoring and other data gathering.[1,2,3,4,5]

**Case 6.1: The Danube River Basin's TransNational Monitoring Network**
The TransNational Monitoring Network (TNMN) is a significant tool operating under the Danube River Protection Convention (DRPC). The DRPC obliges its contracting parties to collaborate in the monitoring and evaluation of the Danube River Basin. The TNMN was officially launched in 1996 with the aim of providing a comprehensive and well-rounded view of pollution, long-term trends in water quality, and pollution loads in the major rivers of the basin. The TNMN was established by the International Commission for the Protection of the Danube River (ICPDR) to support the DRPC in the field of monitoring and evaluation using data assessed at the national level. The DRPC calls for the harmonization of monitoring and evaluation methods, particularly concerning water quality in rivers. The TNMN is tasked with developing coordinated or joint monitoring systems, including stationary or mobile measurement devices, shared communications, and data processing facilities. The TNMN also seeks to elaborate and implement joint programs for monitoring riverine conditions in the Danube catchment area, including flow rates, water quality, sediments, and riverine ecosystems as the basis for the assessment of transboundary impacts. The TNMN's primary objective is to provide a structured and well-balanced overall view of pollution and long-term trends in water quality and pollution loads in the major rivers of the Danube River Basin. The collected data is published annually in the TNMN Yearbooks. In 2006, the TNMN underwent a revision to ensure compliance with the provisions of the European Union Water Framework Directive. The TNMN monitoring network is derived from national surface water monitoring networks and includes 101 monitoring stations with up to three sampling points across the Danube and its main tributaries. The minimum sampling frequency is 12 times per year for chemical determinants in water and twice a year for biological parameters. The TNMN's assessment of loads in the Danube River also provides estimates of the influx of

polluting substances into the Black Sea. This information is vital in supporting policy development. The special load assessment programme began in 2000, calculating pollution loads for Biochemical Oxygen Demand (BOD)5, Inorganic nitrogen, Ortho-phosphate-phosphorus, Dissolved phosphorus, Total phosphorus, Suspended solids, and, on a discretionary basis, chlorides.[6]

## 6.2 Permits

The goal of environmental permitting is to protect human health and the environment by defining, in a transparent, accountable manner, legally binding requirements for individual sources of significant environmental impact. Single-medium permitting, which is the traditional regulatory approach, is based on addressing specific environmental problems such as water protection. Specifically, under this type of regime, the limit for environmental impacts of installations are set to protect the environmental medium (for example, water). Meanwhile, integrated permitting means that emissions to air, water, and land, as well as other environmental effects, must all be considered together. This means that regulators set permit conditions to achieve a high level of protection for the environment overall. In the context of managing water quality, regulatory permits are mainly used to control point sources, including wastewater treatment discharges, industrial waste discharges, and stormwater collection systems. The permits are typically issued by the government and specify discharge levels for pollutants. Point sources may not exceed these permitted levels.[7,8] There are a series of fundamental principles in establishing a permitting system:

–   *Permitting of all stationary sources of significant pollution*: All stationary pollution sources with significant environmental impact should have an environmental permit as a precondition for their operation. The provision of environmental permits to industrial installations is a fundamental element of the regulatory process addressing pollution. A coherent permitting system is also necessary to ensure economic competition remains fair under environmental regulations and that economic development proceeds in a sustainable way

–   *Differentiation of regulatory regimes for major and minor pollution sources*: Major pollution sources should be subject to integrated environmental permitting on a case-by-case basis, where all environmental aspects are considered simultaneously, and that the environment is a disposal route of last resort. Small and medium-sized enterprises should be subject to simplified regulatory regimes as these businesses pose a lower environmental risk and case-by-case permitting would pose a disproportionately heavy burden on them as well as on the regulators

–   *Appropriate permitting authority*: There should be a 'one-stop-shop' system where applicants deal with one designated authority that ensures coordination with all other stakeholders. This increases the consistency and predictability of the permitting process and reduces the administrative burden on both government and industry

- *Public participation and access to information*: The public should be allowed to comment on permit applications before the authority reaches its decision and have access to permit-related information after the permit has been awarded. Regarding consulting the public, it is appropriate to maintain a permit register accessible to the public, where applications and permits are placed, subject to commercial confidentiality
- *Extensive stakeholder involvement*: Permitting requires a transparent process for involving all institutional stakeholders. Stakeholder consultations should be part of both the development of the regulatory framework for permitting (procedures, rules, and guidance) and the permit determination process itself. The permitting authority should also consult other authorities with related responsibilities or interests. Permit registers and interagency electronic networks should be developed to facilitate such coordination
- *Outreach to the regulated community*: Environmental authorities should make substantial effort through trade associations, environmental and industry publications, industry seminars etc. to ensure operators are aware of its obligations under the environmental law. The environmental permitting authority may hold pre-application discussions with the operator before it submits a formal application to clarify relevant requirements
- *Close interaction with Environmental Assessment (EA)*: Both EA and environmental permitting follow legally binding procedures for identifying and analysing significant environmental impacts and making decisions related to economic activity. However, EA applies at an earlier stage of project planning and considers a wide range of alternatives and mitigation measures. As such, EA and permitting should be applied to maximise their effectiveness and avoid overlap. This should be achieved through using EA findings in preparing and evaluating permit applications and include EA recommendations on mitigation measures in permit conditions
- *Clear and enforceable permit requirements*: A permit must contain conditions that are unambiguous and enforceable. The key to simple, effective, and consistent permitting is to base permit conditions on statutory requirements and technical guidance that have been developed in cooperation with all stakeholders and are available to all, including the public[9]

### 6.2.1 Tradable permits

Tradable permits are market-based instruments that provide allowance or permission to engage in an activity. These permits are mainly used to allocate pollution rights, and they can be issued under a trading system. There are two main types of trading systems: cap-and-trade systems and baseline-and-credit systems. In a cap-and-trade system, an upper limit on permits is fixed, and the permits are either auctioned or distributed for free according to specific criteria. Under a baseline-and-credit system, there is no fixed limit on pollution, but polluters that reduce their emissions more than they have to earn credits that they sell to others who need them to comply with regulations that they are subject to. Overall, the use of tradable permits has been made on the following grounds:

- Incentives for abatement cost equalisation
- Positive technological innovation and diffusion impacts
- A high degree of environmental certainty
- Relatively low administrative costs
- Flexibility to address distributional concerns[10,11,12]

In the case of water resources, tradable water pollution rights are where the water management authority establishes the maximum amount of emissions according to the carrying capacity of the ecosystem in question. The total amount of emissions is subdivided into a fixed number of permits or rights to pollute, that can be initially allocated according to past levels of pollution (grandfathering) or by auction. The holders can trade the rights in a secondary permit market.[13] For tradable water pollution rights to be successful, there need to be secure property rights, water rights must be enforceable, and an efficient administrative system must exist to ensure market operation. Overall, tradable water pollution rights systems can provide greater flexibility on the timing and level of technology a facility might install, reduce overall compliance costs, and encourage the voluntary participation of non-point sources within a river basin. Furthermore, trading can provide additional environmental benefits, including carbon sinks, flood retention, riparian improvement, and habitat.[14,15] There are a variety of trading scenarios possible for tradable water pollution rights systems, including:

### 6.2.1.1 Point source-point source trading

Trading between point sources is the most basic form of water quality trading. It is relatively straight forward, easily measurable, and directly enforceable. It is usually the most accessible type of trading to implement, measure reductions from, and ensure compliance and enforcement with because all sources have a permit, the effectiveness of removal technologies is relatively known, and monitoring protocols are in place. Several trading scenarios exist for point source-point source trading, which is summarised in Table 6.1.

### 6.2.1.2 Point source-non-point source trading

Trading between point source buyers and non-point source sellers provides another opportunity to meet water quality standards. In successful point source-non-point source trading schemes, point sources benefit by purchasing credits for required reductions at a lower cost than technology upgrades. Non-point sources benefit by gaining income

**Table 6.1:** Point Source-Point Source Trading Scenarios.

| Scenario | Description |
|---|---|
| Trading between two point sources | – Generally, it involves a trade agreement between two point sources<br>– One point source is the credit generator, and the other is the credit purchaser<br>– A single permit can be issued that incorporates or references the trade agreement and includes both point sources as co-permittees. Alternatively, each discharger can be issued an individual permit with trading provisions placed in each permit |
| Multiple facility point-source trading/No exchange | – Involves a group of point sources operating under a single trade agreement<br>– The agreement can establish ground rules for trading to allow point sources to trade among themselves<br>– The agreement can precisely identify the point sources that may participate in water quality trading, or it can identify a geographic boundary (typically a river basin) or a type of discharger, or both, and allow qualifying point sources to participate in trading as desired or appropriate<br>– An over-all limit or cap set by the permit regulates all trades |
| Point source credit exchanges | – Point sources purchase credits from a central exchange to comply with individual effluent limitations<br>– The credit exchange is likely to be either operated by or approved and overseen by a state regulatory agency<br>– Credits in the exchange are generated by point sources that control their discharges<br>– The trade agreement can specify how credits may be generated and purchased, how trade ratios are calculated, and individual and group responsibilities for meeting effluent limitations and overall pollutant loading caps |

from better resource management, and water quality improves. Several types of trading scenarios exist for point source-non-point source trading, which is summarised in Table 6.2.[16]

**Table 6.2:** Point Source-Non-point Source Trading Scenarios.

| Scenario | Description |
|---|---|
| Single point source-non-point source trades | – This is a trade agreement between a single point source and one or more non-point sources<br>– Under this trade, the non-point source(s) reduce(s) pollutant loads below the established baseline to generate credits, and the point source purchases these credits |

**Table 6.2** (continued)

| Scenario | Description |
| --- | --- |
| Non-point source credit exchange | – A credit exchange programme is established to buy credits from multiple non-point sources to sell to point sources<br>– The exchange could be managed by the state, a conservation district, a private entity, or another third party<br>– A broker can be used to identify trading partners and facilitate trades<br>– There are two main types of exchanges:<br> 1. A broker-facilitated exchange where the broker brings parties together to trade directly with each<br> 2. A central exchange where the point sources are not required to deal directly with non-point sources<br>– For the second type of exchange, the credit sellers (non-point sources) generate pollutant load reductions using a variety of approved BMPs and sell the credits to the credit exchange. Point sources may then purchase credits from the credit exchange rather than directly from the non-point sources |

**Case 6.2: Water Quality Credit Trading in the Great Miami River Watershed**
Over the past three decades, the Great Miami River Watershed in southwest Ohio experienced marked improvements in surface water quality. However, approximately 40 percent of the watershed's rivers and streams, primarily in the headwaters areas, continued to fail to meet water quality standards, with excess nutrients contributing to this failure locally and downstream in the Gulf of Mexico. Failure to attain these standards triggered additional regulations focused on wastewater treatment plants, but because over 70% of the land in the watershed was used for agriculture, nutrient-related water quality challenges primarily related to agricultural land use. Although agricultural producers in the watershed implemented conservation farming practices, available federal, state, and local incentives did not match the needs. To address this issue, the Miami Conservancy District collaborated with federal, state, and local partners to design and implement a market-based program that reduced nutrients in streams and rivers as an alternative to traditional regulatory strategies. The Great Miami River Watershed Water Quality Credit Trading Program (Trading Program) was launched in 2004 as a pilot to evaluate the viability of water quality credit trading as an approach to reduce nutrients in the watershed. The program established a new sustainable local source of revenue for agricultural producers to implement conservation practices in cooperation with wastewater treatment plants. The Trading Program continued in pilot much longer than expected because anticipated nutrient regulations were still not in place more than ten years after they were originally announced by Ohio EPA. To fund the pilot, over $3 million in funding was provided by wastewater treatment plants, the Ohio Department of Natural Resources, the U.S. Department of Agriculture, and the U.S. Environmental Protection Agency. An extensive economic and market analysis conducted prior to Trading Program design estimated that the cost for implementation of agricultural conservation practices to achieve a similar level of nutrient reduction was projected at $37.8 million, a potential $384.7 million savings compared to wastewater treatment plant upgrades.[17]

## 6.3 Best management practices

River basin management strategies generally involve controlling non-point source pollution by implementing various BMPs.[18]

### 6.3.1 Agricultural best management practices for water quality protection

There are a variety of agricultural BMPs that can protect water quality, including the following.

#### 6.3.1.1 Conservation tillage
Low-till agriculture also known as conservation or reduced till consists of a combination of a crop harvest that leaves at least 30 percent of the soil surface covered after planting. This slows water movement, which reduces the amount of soil erosion and potentially leads to greater infiltration. Meanwhile, no-till farming is a type of conservation tillage that seeks to minimise soil disruption and entails leaving crop residue on the fields after harvest. The residue acts as a mulch to stabilise and protect the soil from wind and water erosion. Leaving crop residues on the soil surface can also increase water infiltration by helping slow and capture runoff, which in turn helps conserve water and enhances the utilisation of applied fertilisers and pesticides. By reducing the amount of surface runoff, conservation tillage can help reduce contamination of nearby water bodies by reducing the transport of sediment, fertilisers, and pesticides.[19,20,21]

#### 6.3.1.2 Crop nutrient management
To successfully grow and produce crops, plants must receive sufficient and proper nutrients at correct times and in appropriate amounts. The practice of nutrient management involves effectively managing the amount, source, placement, form, and timing of the application of plant nutrients and soil amendments. Nutrient management not only helps retain optimum production levels but can also protect water quality and reduce input costs. By carefully managing nutrients and preventing overapplication of fertilisers and manure, the amount of excess nutrients lost to runoff can be reduced, which in turn reduces the amount of non-point source pollution from cropland.[22]

#### 6.3.1.3 Riparian buffers
Riparian buffers are areas of trees or other vegetation located adjacent to a water body and are managed to reduce the negative impact of nearby land use. They have a variety of roles including separating the crop field from the stream, filtering runoff to remove sediment, nutrients, pesticides, and microorganisms, increasing water infil-

tration, taking up nitrate from shallow groundwater, and stabilising streambanks. Riparian buffers reduce nitrogen loading into the stream by filtering and sedimentation of organic and other particulate-bound nitrogen. In particular, riparian buffers increase infiltration, increase nitrogen uptake, especially if there is subsurface flow through the root zone, and increase denitrification, with denitrification relatively high with mature riparian forest, intermediate with a grass buffer, and least with cropland.[23,24,25]

### 6.3.1.4 Irrigation water management

Irrigation scheduling involves the application of irrigation water based on systematic monitoring of crop soil-moisture requirements. It should be based on the daily water use of the crop, the water-holding capacity of the soil, the lower limit of soil moisture for the crop, and the volume of water applied to the field. Irrigation water should be applied in a way that ensures efficient use and distribution, and that minimises runoff, deep percolation, and soil erosion. When chemigation, or the application of fertilisers, pesticides, or other chemicals through irrigation water, is implemented, precautions should be taken to prevent chemigated water from contaminating surface or groundwater. Irrigation practices that can minimise surface and groundwater contamination include:

- Knowing the water-holding capacity of all soils in the field
- Monitoring soil water content to determine how much water has been removed and to evaluate the effectiveness of current irrigation management practices
- Recording how much water is being delivered to the field
- Recording precipitation and estimating how much enters the crop root zone
- Estimating crop water use for each crop
- Calculating a soil water balance based on stored soil water, crop water use, and water applied via precipitation or irrigation[26,27]

**Case 6.3: Virginia's Ag BMP Loan Program**
The Virginia Clean Water State Revolving Loan Fund (VCWRLF) is a state financing assistance program that is managed by the Virginia Resources Authority (VRA) and administered by the Virginia Department of Environmental Quality (DEQ) on behalf of the State Water Control Board (SWCB). One of the components of the VCWRLF is the Ag BMP Loan Program, which provides low-cost financing to Virginia's agricultural producers and Soil and Water Conservation Districts (SWCDs) to reduce agricultural non-point source pollution. From 2000 to 2016, the Ag BMP Loan Program provided $46 million in 491 loans for 615 Ag BMP projects throughout Virginia. The program resumed on July 1, 2019, with additional incentives provided to applicants. The loan program offers loans at 0% interest, with the potential for principal forgiveness and no long-term loan requirement. The program has expanded the list of eligible practices and applicants. Disbursements of loan proceeds are made on a reimbursement basis, and debt repayment begins six months after project completion. Repayment schedules range from one to ten years, depending on the loan amount and asset useful life. Eligible practices for the loan program include surface water and groundwater protection practices, stream fencing and alternative watering systems, tree plantings along streams, stormwater impoundment and sediment retention, streambank stabilization and stream crossings, water table control structure and constructed

wetlands, sinkhole protection, livestock and structural practices, roof runoff management systems, animal waste structures and pumping systems, manure and animal composting systems, animal mortality incinerators, chemical and fertilizer handling facilities, relocation of confined feeding operations, grazing systems, animal travel lanes, loafing lots, and no-till drill equipment. The loan program has a minimum allowable loan amount of $10,000, with the VRA as the minimum loan amount provider. For projects that have received cost share funds before loan closing, loans may be less than $10,000, provided that the total project cost, including the cost share amount, is at least $10,000. The maximum allowable loan amount is $600,000, and the maximum amount of active program loan utilization is $1,000,000. Active program loan utilization is the current loan amount outstanding in the program and the amount of any new loan application(s).[28]

### 6.3.2 Industrial best management practices for water quality protection

A variety of BMPs can be designed to prevent or reduce the effects of pollutants on waterways and habitat health from industrial stormwater discharges:

#### 6.3.2.1 Best management practices to treat total suspended solids
Total suspended solids (TSS) is inorganic (for example, sand, metals) and organic (for example, vegetative and animal waste) particles and debris that are washed off surfaces into surface waters. TSS smothers fish eggs and larvae, causes turbidity that impairs sight-feeding fish, increases drinking water treatment costs, and acts as a vehicle to transport other pollutants such as nutrients and metals to surface waters. Control of TSS can be achieved by avoiding or minimising land disturbance such as clearing and grading as well as reducing stormwater runoff from construction, mining, and logging sites. BMPs that reduce TSS include:
- Implementing a frequent outdoor sweeping schedule
- Using grassed or vegetated areas to catch sediment particles in flowing stormwater, for example, adding rock filters upstream of existing grassed areas to slow down water velocity and adding fibre or synthetic mats on eroded areas or bare, non-vegetation areas
- Developing detention ponds[29]

#### 6.3.2.2 Best management practices to manage availability of oxygen
The availability of oxygen for aquatic life is impacted by substances in water which use oxygen to break down materials washed into the water. Biological oxygen demand (BOD) is the amount of oxygen needed by aerobic biological organisms in a body of water to break down the organic material present. Chemical oxygen demand (COD) measures the oxygen required to decompose organic and inorganic materials, metals, and nutrients present in water by chemical reaction. The higher the BOD or

COD, the less oxygen is available in the water for fish and other aquatic life. A variety of BMPs can be used to manage BOD and COD, including:
- Erosion control
- Litter prevention/management
- Stormwater detention ponds
- Constructed wetlands
- Filtration devices
- Infiltration devices[30]

### 6.3.2.3 Best management practices to manage nutrients

Excessive amounts of some nutrients can lead to algal blooms or other conditions toxic to aquatic life and detrimental to human health:

High amounts of phosphorus can create algal blooms and excessive plant growth which results in oxygen depletion and accelerated sediment filling of lakes when the algae die. Phosphorus sources include chemicals and fertilisers, animal wastes and by-products, food/energy processing wastes, wood processing wastes, and cleaning agents. BMPs used to treat phosphorus include:
- Properly storing materials
- Cleaning up materials from impervious surfaces
- Covering raw material, waste piles, and transfer processes
- Storing materials indoors or covering with a roof or tarp
- Capturing and treating high-strength waste streams separately
- Slowing down water to allow nutrient attenuation by grasses/vegetation before runoff occurs

Nitrates and nitrogen-containing substances can affect both surface water and groundwater with large concentrations of nitrates presenting a health hazard in groundwater and drinking water. Nitrate/nitrogen sources include fertiliser manufacturing, mining, food manufacturing, leather tanning, and fabricated metal manufacturing activities. BMPs used to treat nitrates/nitrogen include:
- Source control by implementing fertiliser application limits
- Minimising, or eliminating exposure before discharge
- Housekeeping such as sweeping spilt solid materials, and detention ponds[31]

### 6.3.2.4 Best management practices to manage metals

Metals originate from galvanising, chrome plating, and other industrial operations. As metals corrode, dissolve, or settle out in the air, small amounts are carried away by the wind or water and can concentrate in stormwater runoff. Many of these metals become attached to sediment particles and are carried with it to receiving waters.

When these sediments settle out, the attached metals accumulate over time to concentrations that are harmful to sediment-dwelling and other aquatic life. BMPs used to treat metals include:

- Source control by limiting metal exposure to stormwater
- Modifying processes, storage, or handling
- Minimising or eliminating the usage of metal-containing product processes
- Replacing or painting galvanised surfaces
- Implementing vegetative buffer strips to capture sediment particles
- Adding recycling to recover and recycle specific metals from the production processes[32]

---

**Case 6.4: California's Industrial & Commercial BMP Online Handbook**

The Industrial & Commercial BMP Online Handbook is an online reference tool designed to help industrial businesses comply with California stormwater regulations. It contains monitoring guidance and inspection forms, including a Storm Water Pollution Prevention Plan (SWPPP) Template. The latest version of the handbook (December 2019) and SWPPP Template are included in a subscription to the Industrial & Commercial BMP Online Handbook, which provides regular updates at least annually. The new Water Quality Based Corrective Action Report Template assists dischargers who trigger a WQBCA by providing guidance for facility evaluation, BMP assessment, SWPPP assessment, and documentation submission in the California Storm Water Multiple Application and Report Tracking System. Updates to the handbook in 2019 include the addition of persistent organic pollutants to the list of potential pollutants, conditions, and impacts, a separate regulatory definition of trash, updated Decision Trees with Non-Exposure Coverage, expanded definitions and requirements for minimum and advanced BMPs, updated SWPPP signature requirements, alphanumeric designations to distinguish maintenance sheets from fact sheets and guide sheets, and the addition of Permeable Pavement and Green Roof fact sheets.[33]

---

### 6.3.3 Urban best management practices for water quality protection

In urban settings, surfaces are subject to the deposit of contaminants, which are then subject to wash-off by rainfall or snowmelt. The typical contributors to pollutants in runoff include vehicular traffic, lawn care, pets, eroded sediments, and vegetative litter. The major urban non-point source pollutants include sediment, nutrients, oxygen-demanding substances, toxic chemicals, chloride, bacteria and viruses, and temperature changes. Stormwater BMPs aim to prevent or reduce the movement of sediment, nutrients, pollutants, or debris from land to surface or ground waters. There are a variety of structural and non-structure BMPs available that are summarised in Table 6.3.[34,35,36]

**Table 6.3:** Structural and Non-Structural Stormwater Best Management Practices.

| Type | Best Management Practice | Description |
| --- | --- | --- |
| Structural | Wet extended detention ponds | These are a combination of permanent pool storage and extended detention storage above the permanent pool to provide additional water quality or rate control |
| | Dry ponds | These have no permanent pool. Instead, they rely on extended detention storage for treatment. They are best combined with other BMPs such as filtration or infiltration |
| | Stormwater wetlands (constructed wetlands) | Stormwater wetlands are constructed management practices. They are similar in design to stormwater ponds; however, they differ by varying water depths and associated vegetation. They are typically installed at the downstream end of the treatment train |
| | Infiltration basins | These basins capture and temporarily store stormwater runoff while allowing it to infiltrate into subsurface soils |
| | Underground infiltration devices | These capture stormwater underground in pipes or tanks and allow stormwater to infiltrate into subsurface soils through open bottoms or perforations. They are typically installed where space is limited |
| | Filtration systems | These systems are a diverse group of techniques for treating stormwater runoff. Options can range from a simple sand filter or vegetative filter system through to complex engineered systems. The commonality is that each type utilises one or more forms of media, such as sand, gravel, peat, grass, soil or compost or synthetic media to filter stormwater pollutants |
| Non-structural | Information and education | Erosion control information, fertiliser and pesticide application guides, illicit dumping and littering information, landscaping information to reduce runoff, information on the correct disposal of hazardous waste and used motor oils |
| | Ordinances and regulations | Erosion-control ordinances, comprehensive management plans for developments, elimination of illegal connections, fertiliser and pesticide licensing, land-use controls, landscaping requirements to reduce runoff, special commercial or industrial requirements |
| | Source controls by the City | Limiting infiltration to storm sewers, effective use of de-icing chemicals, management of hazardous waste and used motor oils, monitoring programmes, spill response and prevention, street cleaning, storm sewer maintenance |

**Case 6.5: Melbourne Water's Water Smart City Model**
Melbourne Water has developed the Water Smart City Model using Lego to raise awareness of the benefits of BMPs in managing stormwater quality and surface runoff. The model is an educational activity suitable for all ages that can be used at community events and festivals. The activity involves the audience building a model city with roads and buildings made from Lego building blocks. Food dye, representing pollutants, is placed on the City and rainfall is simulated over the model, carrying the pollution over the impervious surfaces and into the 'bay'. A variety of features including rain gardens, rainwater tanks, swales, and green roofs are then added. Pollution is again added to the model and rain simulated. The amount of surface runoff is significantly decreased due to the retention capabilities of the new features, reducing risks of flooding. Pollution is also captured in the features, so the water flowing into the 'bay' is cleaner.[37]

## 6.4 Source water protection

Source water refers to sources of water, such as rivers, streams, lakes, reservoirs, springs, and groundwater, that provide water to public drinking water supplies and private wells. Protecting the source can reduce risks by preventing exposures to contaminated water. Protecting source water from contamination also helps reduce treatment costs and may avoid or defer the need for complex treatment. Source water protection includes a variety of actions and activities that aim to safeguard, maintain, or improve the quality and/or quantity of sources for drinking water and their contributing areas. Examples of source water protection include:

– Riparian zone restoration
– Streambank stabilisation
– Land protection/conservation easements
– BMPs for agriculture and forestry activities or stormwater control
– Local ordinances to limit certain activities in source water or wellhead protection areas
– Developing emergency response plans
– Educating local industry, businesses, and communities on pollution prevention and source water protection[38,39,40,41]

**Case 6.6: Sustainable Water Management for Drinking Water Protection in Munich, Germany**
Stadtwerke München (SWM) aims to protect the drinking water quality in the Mangfalltal, Loisachtal, and Munich Schotterebene (gravel plain) of Munich. To maintain the purity of spring and groundwater, SWM collaborates with the local population and implements sustainable initiatives such as organic farming, eco-friendly forestry, and habitat preservation. One of SWM's initiatives is the "Ökobauern" programme, which was launched in 1992 to promote organic farming in the Mangfalltal water catchment area. The programme supports farmers in transitioning from traditional to organic farming by partnering with established organic farming associations. Currently, 175 farmers have converted their farms to ecological standards, cultivating an area of approximately 4,400 hectares, which represents one of the largest organically farmed areas in Germany. SWM designates water protection areas

around the sites of all extraction plants to preserve the habitat and diversity of plant and animal species. The natural and water-conserving management of these areas ensures that the water quality is secured. Additionally, the SWM forestry programme ensures the eco-friendly management of the forests in the Mangfalltal and Munich Schotterebene. A healthy forest acts as an ideal water reservoir and provides a species-appropriate habitat for numerous animals.[42]

# Notes

1 New York State Department of State, "New York State Guidebook Watershed Plans Protecting and Restoring Water Quality," (2009), https://www.dos.ny.gov/opd/sser/pdf/WatershedPlansGuidebook.pdf.

2 Texas A&M University, "Watershed Approach to Water Quality Managment," https://texaswater.tamu.edu/surface-water/watershed-water-quality-management.html.

3 A. Said et al., "Exploring an Innovative Watershed Management Approach: From Feasibility to Sustainability," *Energy* 31, no. 13 (2006).

4 Charalampos Skoulikaris and Antigoni Zafirakou, "River Basin Management Plans as a Tool for Sustainable Transboundary River Basins' Management," *Environmental Science and Pollution Research* 26, no. 15 (2019).

5 Marta Terrado et al., "Integrating Ecosystem Services in River Basin Management Plans," *Journal of Applied Ecology* 53, no. 3 (2016).

6 ICPDR, "Tnmn – Transnational Monitoring Network," https://www.icpdr.org/main/activities-projects/tnmn-transnational-monitoring-network.

7 Texas A&M University, "Watershed Approach to Water Quality Managment".

8 World Bank, "Watershed Management Approaches, Policies, and Operations: Lessons for Scaling Up," (2008), http://siteresources.worldbank.org/TURKEYEXTN/Resources/361711-1216301653427/5218036-1267432900822/WatershedExperience-en.pdf

9 OECD, "Guiding Principles of Effective Environmental Permitting Systems" (2007), https://www.oecd.org/env/outreach/37311624.pdf

10 "Oecd Policy Instruments for the Environment," (2016), http://www.oecd.org/environment/tools-evaluation/PINE_Metadata_Definitions_2016.pdf

11 "Tradeable Permits: Policy Evaluation, Design and Reform," (2004), https://www.oecd-ilibrary.org/environment/tradeable-permits_9789264015036-en.

12 "Efficient and Effective Use of Tradeable Permits in Combination with Other Policy Instruments" (2003), www.oecd.org/env/cc/2957650.pdf

13 Simone Borghesi, "Water Tradable Permits: A Review of Theoretical and Case Studies," *Journal of Environmental Planning and Management* 57, no. 9 (2014).

14 SSWM, "Tradable Water Rights," http://archive.sswm.info/category/implementation-tools/water-distribution/software/economic-tools/tradable-water-rights.

15 US EPA, "Water Quality Trading," https://www.epa.gov/npdes/water-quality-trading.

16 "Water Quality Trading Toolkit for Permit Writers," (2009), https://www3.epa.gov/npdes/pubs/wqtradingtoolkit.pdf

17 Miami Conservancy District, "Water Quality Credit Trading Program: A Common Sense Approach to Reducing Nutrients," (2014), https://www.mcdwater.org/wp-content/uploads/PDFs/2014WQCTPfactsheet.pdf.

18 Jing Wu, Shaw L. Yu, and Rui Zou, "A Water Quality-Based Approach for Watershed Wide Bmp Strategies1," *JAWRA Journal of the American Water Resources Association* 42, no. 5 (2006).

19 Natural Water Retention Measures, "Low Till Agriculture," (2015), http://nwrm.eu/measure/low-till-agriculture.

**20** Wyoming Department of Environmental Quality Water Quality Division Nonpoint Source Program, "Cropland Best Management Practice Manual: Conservation Practices to Protect Surface and Ground Water," (2013), http://deq.wyoming.gov/media/attachments/Water%20Quality/Nonpoint%20Source/Best%20Management%20Practices/2013_wqd-wpp-Nonpoint-Source_Cropland-Best-Management-Practice-Manual.pdf.

**21** Steffen Seitz et al., "Conservation Tillage and Organic Farming Reduce Soil Erosion," *Agronomy for Sustainable Development* 39, no. 1 (2018).

**22** Wyoming Department of Environmental Quality Water Quality Division Nonpoint Source Program, "Cropland Best Management Practice Manual: Conservation Practices to Protect Surface and Ground Water".

**23** Ibid.

**24** Matt Helmers and Antonio Mallarino, "Agricultural Phosphorus Management and Water Quality Protection in the Midwest," (2005), https://store.extension.iastate.edu/product/Agricultural-Phosphorus-Management-and-Water-Quality-Protection-in-the-Midwest-EPA-Region-7.

**25** Venkatachalam Anbumozhi, Jay Radhakrishnan, and Eiji Yamaji, "Impact of Riparian Buffer Zones on Water Quality and Associated Management Considerations," *Ecological Engineering* 24, no. 5 (2005).

**26** Wyoming Department of Environmental Quality Water Quality Division Nonpoint Source Program, "Cropland Best Management Practice Manual: Conservation Practices to Protect Surface and Ground Water".

**27** Mallarino, "Agricultural Phosphorus Management and Water Quality Protection in the Midwest".

**28** Virginia Department of Environmental Quality, "Virginia Clean Water Revolving Loan Fund (Vcwrlf)," https://www.deq.virginia.gov/our-programs/water/clean-water-financing-and-assistance/virginia-clean-water-revolving-loan-fund-vcwrlf.

**29** Minnesota Pollution Control Agency, "Industrial Stormwater: Best Management Practices Guidebook," (2015), https://www.pca.state.mn.us/sites/default/files/wq-strm3-26.pdf

**30** Ibid.

**31** Ibid.

**32** Ibid.

**33** California Stormwater Quality Association, "About the Industrial & Commercial Bmp Handbook," https://www.casqa.org/resources/bmp-handbooks/industrial-commercial-bmp.

**34** Mahesh R. Gautam, Kumud Acharya, and Mark Stone, "Best Management Practices for Stormwater Management in the Desert Southwest," *Journal of Contemporary Water Research & Education* 146, no. 1 (2010).

**35** Minnesota Pollution Control Agency, "Industrial Stormwater: Best Management Practices Guidebook".

**36** R.C. Brears, *Blue and Green Cities: The Role of Blue-Green Infrastructure in Managing Urban Water Resources* (Palgrave Macmillan UK, 2018).

**37** "The Role of Blue-Green Infrastructure in Managing Urban Water Resources," Mark and Focus https://medium.com/mark-and-focus/the-role-of-blue-green-infrastructure-in-managing-urban-water-resources-dd058007ba1a

**38** US EPA, "Source Water Protection," https://www.epa.gov/sourcewaterprotection.

**39** The Forest Guild and Mountain Conservation Trust of Georgia American Rivers, "Forests to Faucets: Protecting Upstream Forests for Clean Water Downstream," (2013), https://americanrivers.org/wp-content/uploads/2016/05/AmericanRivers_forests-to-faucets-report.pdf.

**40** G. Tracy Mehan III and Adam T. Carpenter, "Bringing Agriculture and Drinking Water Utilities Together for Source Water Protection," *Journal – AWWA* 111, no. 8 (2019).

**41**  Monica B. Emelko et al., "Implications of Land Disturbance on Drinking Water Treatability in a Changing Climate: Demonstrating the Need for "Source Water Supply and Protection" Strategies," *Water Research* 45, no. 2 (2011).

**42**  SWM, "M/Water," https://www.swm.de/english/m-wasser.

# References

American Rivers, The Forest Guild and Mountain Conservation Trust of Georgia. "Forests to Faucets: Protecting Upstream Forests for Clean Water Downstream". (2013). https://americanrivers.org/wp-content/uploads/2016/05/AmericanRivers_forests-to-faucets-report.pdf.

Anbumozhi, Venkatachalam, Jay Radhakrishnan, and Eiji Yamaji. "Impact of Riparian Buffer Zones on Water Quality and Associated Management Considerations". *Ecological Engineering* 24, no. 5 (2005/05/30/ 2005): 517–23.

Borghesi, Simone. "Water Tradable Permits: A Review of Theoretical and Case Studies". *Journal of Environmental Planning and Management* 57, no. 9 (2014/09/02 2014): 1305–32.

Brears, R.C. *Blue and Green Cities: The Role of Blue-Green Infrastructure in Managing Urban Water Resources.* Palgrave Macmillan UK, 2018.

_____. "The Role of Blue-Green Infrastructure in Managing Urban Water Resources". Mark and Focus https://medium.com/mark-and-focus/the-role-of-blue-green-infrastructure-in-managing-urban-water-resources-dd058007ba1a

City of Guelph. "Stormwater Service Credits for Business". https://guelph.ca/living/environment/water/stormwater/stormwater-service-fee-credit-program/

Emelko, Monica B., Uldis Silins, Kevin D. Bladon, and Micheal Stone. "Implications of Land Disturbance on Drinking Water Treatability in a Changing Climate: Demonstrating the Need for "Source Water Supply and Protection" Strategies". *Water Research* 45, no. 2 (2011/01/01/ 2011): 461–72.

First Climate. "Water Quality Credits". https://www.firstclimate.com/en/water-quality-credits/

Gautam, Mahesh R., Kumud Acharya, and Mark Stone. "Best Management Practices for Stormwater Management in the Desert Southwest". *Journal of Contemporary Water Research & Education* 146, no. 1 (2010/12/01 2010): 39–49.

Government of Saskatchewan. "Farm Stewardship Program (Fsp)". https://www.saskatchewan.ca/business/agriculture-natural-resources-and-industry/agribusiness-farmers-and-ranchers/canadian-agricultural-partnership-cap/environmental-sustainability-and-climate-change/farm-stewardship-program-fsp

ICPDR. "Tnmn – Transnational Monitoring Network". https://www.icpdr.org/main/activities-projects/tnmn-transnational-monitoring-network.

_____. "Watching the Danube´ Beyond the Jds" (2013). http://www.danubesurvey.org/jds3/jds3-files/nodes/documents/factsheet7-jds3_1.pdf.

Mallarino, Matt Helmers and Antonio. "Agricultural Phosphorus Management and Water Quality Protection in the Midwest". (2005). https://store.extension.iastate.edu/product/Agricultural-Phosphorus-Management-and-Water-Quality-Protection-in-the-Midwest-EPA-Region-7.

Mehan III, G. Tracy, and Adam T. Carpenter. "Bringing Agriculture and Drinking Water Utilities Together for Source Water Protection". *Journal – AWWA* 111, no. 8 (2019): 34–39.

Minnesota Pollution Control Agency. "Industrial Stormwater: Best Management Practices Guidebook". (2015). https://www.pca.state.mn.us/sites/default/files/wq-strm3-26.pdf

Natural Water Retention Measures. "Low Till Agriculture". (2015). http://nwrm.eu/measure/low-till-agriculture.

New York State Department of State. "New York State Guidebook Watershed Plans Protecting and Restoring Water Quality". (2009). https://www.dos.ny.gov/opd/sser/pdf/WatershedPlansGuidebook.pdf.

NYC DEP. "High Quality Nyc Tap Water Receives New Filtration Waiver". https://www1.nyc.gov/office-of-the-mayor/news/779-17/high-quality-nyc-tap-water-receives-new-filtration-waiver

OECD. "Efficient and Effective Use of Tradeable Permits in Combination with Other Policy Instruments" (2003). www.oecd.org/env/cc/2957650.pdf

____. "Guiding Principles of Effective Environmental Permitting Systems" (2007). https://www.oecd.org/env/outreach/37311624.pdf

____. "Oecd Policy Instruments for the Environment". (2016). http://www.oecd.org/environment/tools-evaluation/PINE_Metadata_Definitions_2016.pdf

____. "Tradeable Permits: Policy Evaluation, Design and Reform". (2004). https://www.oecd-ilibrary.org/environment/tradeable-permits_9789264015036-en.

Said, A., G. Sehlke, D. K. Stevens, T. Glover, D. Sorensen, W. Walker, and T. Hardy. "Exploring an Innovative Watershed Management Approach: From Feasibility to Sustainability". *Energy* 31, no. 13 (2006/10/01/2006): 2373–86.

Seitz, Steffen, Philipp Goebes, Viviana Loaiza Puerta, Engil Isadora Pujol Pereira, Raphaël Wittwer, Johan Six, Marcel G. A. van der Heijden, and Thomas Scholten. "Conservation Tillage and Organic Farming Reduce Soil Erosion". *Agronomy for Sustainable Development* 39, no. 1 (2018/12/18 2018): 4.

Skoulikaris, Charalampos, and Antigoni Zafirakou. "River Basin Management Plans as a Tool for Sustainable Transboundary River Basins' Management". *Environmental Science and Pollution Research* 26, no. 15 (2019/05/01 2019): 14835–48.

SSWM. "Tradable Water Rights". http://archive.sswm.info/category/implementation-tools/water-distribution/software/economic-tools/tradable-water-rights.

Terrado, Marta, Andrea Momblanch, Mònica Bardina, Laurie Boithias, Antoni Munné, Sergi Sabater, Abel Solera, and Vicenç Acuña. "Integrating Ecosystem Services in River Basin Management Plans". *Journal of Applied Ecology* 53, no. 3 (2016): 865–75.

Texas A&M University. "Watershed Approach to Water Quality Managment". https://texaswater.tamu.edu/surface-water/watershed-water-quality-management.html.

US EPA. "Source Water Protection". https://www.epa.gov/sourcewaterprotection.

____. "Water Quality Trading". https://www.epa.gov/npdes/water-quality-trading.

____. "Water Quality Trading Toolkit for Permit Writers". (2009). https://www3.epa.gov/npdes/pubs/wqtradingtoolkit.pdf

Wyoming Department of Environmental Quality Water Quality Division Nonpoint Source Program. "Cropland Best Management Practice Manual: Conservation Practices to Protect Surface and Ground Water". (2013). http://deq.wyoming.gov/media/attachments/Water%20Quality/Nonpoint%20Source/Best%20Management%20Practices/2013_wqd-wpp-Nonpoint-Source_Cropland-Best-Management-Practice-Manual.pdf.

California Stormwater Quality Association. "About the Industrial & Commercial Bmp Handbook." https://www.casqa.org/resources/bmp-handbooks/industrial-commercial-bmp.

ICPDR. "Tnmn – Transnational Monitoring Network." https://www.icpdr.org/main/activities-projects/tnmn-transnational-monitoring-network.

Miami Conservancy District. "Water Quality Credit Trading Program: A Common Sense Approach to Reducing Nutrients." (2014). https://www.mcdwater.org/wp-content/uploads/PDFs/2014WQCTPfactsheet.pdf.

SWM. "M/Water." https://www.swm.de/english/m-wasser.

Virginia Department of Environmental Quality. "Virginia Clean Water Revolving Loan Fund (Vcwrlf)." https://www.deq.virginia.gov/our-programs/water/clean-water-financing-and-assistance/virginia-clean-water-revolving-loan-fund-vcwrlf.

SWM. "M/Water." https://www.swm.de/english/m-wasser.

Virginia Department of Environmental Quality. "Virginia Clean Water Revolving Loan Fund (Vcwrlf)." https://www.deq.virginia.gov/our-programs/water/clean-water-financing-and-assistance/virginia-clean-water-revolving-loan-fund-vcwrlf.

# Chapter 7
# Smart digital water management and managing customers of the future

**Abstract:** Smart digital water management is the use of Information and Communication Technology to provide real-time, automated data for use in resolving water challenges across a range of scales and differing contexts. Smart digital water management enables water utilities and customers to integrate smart principles into their strategies. Meanwhile, water utilities need to move away from viewing customers as recipients of services and instead view them as active participants in the delivery of those services. At the same time, there are growing customer expectations of the level of service delivered by water utilities.

**Keywords:** River Basin Management, Tradable Permits, Water Quality, Best Management Practices

## Introduction

Smart digital water management is the use of Information and Communication Technology (ICT) to provide real-time, automated data for use in resolving water challenges across a range of scales and differing contexts. Smart digital water management enables water utilities to integrate smart principles into urban, regional, and national strategies. Meanwhile, customers are not just passive consumers of water services, but instead, they are in the middle of the water chain, with customer behaviour demanding clean water, which affects the volume of water taken from the environment, treated, and transported for use. At the same time, it is the customer's behaviour that drives demand for how much wastewater needs to be removed, treated, and returned to the environment. Furthermore, customers are becoming active participants in the delivery of water resources while expecting higher levels of service.[1] This chapter will first discuss the concept of smart digital water management, followed by its components before discussing managing customers of the future.

## 7.1 Smart digital water management

There are many applications for smart digital water management including water quality monitoring, water efficiency improvement, efficient irrigation, leak detection, pressure and flow management, and floods and drought monitoring.[2,3] Smart digital water management allows conventional water and wastewater systems to become:
- *Instrumented*: The ability to detect, sense, measure, and record data

https://doi.org/10.1515/9783111028101-007

- *Interconnected*: The ability to communicate and interact with operators and people who manage the systems
- *Intelligent*: The ability to analyse the situation, enable quick responses, and optimise troubleshooting solutions

Smart systems allow informed and systematic decision making for water utilities, based on accurate and timely information. There are a few benefits that can be realised by implementing smart digital water management, examples of which are summarised in Table 7.1.[4,5]

**Table 7.1:** Benefits of Smart Digital Water Management.

| Benefit | Description |
| --- | --- |
| Social | - Improved access to clean water and sanitation through water treatment and monitoring<br>- Health improvements through access to clean, safe water<br>- Improved livelihoods through job creation, higher productivity, and educational opportunities<br>- Greater collaboration with community through increased engagement and knowledge-sharing<br>- Increased gender equality through increased opportunities for capacity building and further education |
| Economic | - Increased efficiency in water and wastewater treatment systems<br>- Reduced waste through lower leakage levels<br>- Job creation through project research, design, development, and implementation<br>- Reduction in future infrastructure costs by improving capacity and efficiency, resulting in less need for additional infrastructure |
| Environmental | - Improved water quality through reduced pollution and contamination of waterways<br>- Improved ecosystem health and protection through improved water quality and water quantity<br>- Increased groundwater protection<br>- Lower carbon emissions from reduced energy consumption and increased energy efficiency<br>- Lower water consumption through leak detection and reduced demand |

## 7.1.1 Categories of smart digital water management technologies

Smart digital water management technologies can be divided into three categories depending on who is using or adopting the technology:

- *Type 1 institutional user*: Technologies are aimed at major institutional users such as water suppliers, water managers, and water treatment plants. Users adopt the technologies in a straightforward manner due to incentives, for example, im-

proved efficiency, environmental benefits or because of regulations or targets introduced by government agencies
- *Type 2 individual user*: Technologies are aimed at many individual users. The implementation of technologies is more complex as it requires individuals to change what they are doing. Individuals are less likely to respond to economic incentives because of perceived inconveniences of taking up the new technology. However, the total impact is large and therefore social benefit is high
- *Type 3 institutional and individuals combined*: This is where an institution develops and implements a technology, but the success relies on the individual user, and therefore this approach requires some engagement[6]

### 7.1.2 Smart digital water management system components

Smart digital water management system components can be divided into digital output instruments (meters and sensors), Supervisory Control and Data Acquisition (SCADA) systems, Geographic Information Systems (GIS), and software for a wide range of purposes.

#### 7.1.2.1 Digital output instruments
Digital output instruments are used to collect and transmit information in real-time for a variety of applications including water quality monitoring, real time leak detection, and pressure management:
- *Water quality monitoring*: Water quality is impacted by non-point and point sources and includes sewage discharge, discharge from industries, runoff from agricultural fields, and urban runoff from impervious surfaces. Other sources of contamination include floods and droughts. Water quality monitoring is the collection of information at set locations at regular intervals that provide data that can be used to determine current conditions and establish trends. The main objectives of online water quality monitoring include measurement of key water quality parameters including microbial, physical, and chemical properties as to identify deviations in parameters and provide early warning of hazards. Also, real-time monitoring can inform stakeholders of activities impacting water quality
- *Real-time leak detection*: Meters are usually read by manual meter reading which is an expensive and highly labour-intensive job. Water leaks can go undetected for long periods using manual meters, resulting in mounting damage, and wasted water. Smart water meters can monitor and detect leaks based on abnormal flow patterns, especially continuous water flow. Water utilities can install smart meters to detect and quantify the water losses in District Metered Areas (DMA). The water supplied to a DMA can be compared to the consumption volumes during a defined period and a water balance developed

– *Pressure management*: Pressure management is the practice of managing water distribution network pressures to the optimum levels of service, ensuring a sufficient and efficient supply of water to customers. It is one of the most cost-effective ways of reducing leakage in water distribution networks. The objectives of pressure management for reducing leakage is reducing background leakage, which is acoustically undetectable seeps at pipe joints and small cracks and are uneconomical to be repaired on an individual basis, reducing the rate of new leaks and breaks, which occur on mains and service connections, and reducing the flow rate from any leaks and breaks[7,8,9,10,11,12]

**Case 7.1: Real-time water quality monitoring in upstate New York reservoirs**
The New York City Department of Environmental Protection (NYCDEP) sought to upgrade their water quality monitoring system to collect real-time data and improve management and reporting. The manual measurements taken once a day by technicians in remote data locations were deemed insufficient and required an improved system for accurate monitoring. To meet this objective, NYCDEP collaborated with YSI, a company that designs and manufactures water quality monitoring equipment. YSI developed a buoy-based system that continuously monitors water quality in two upstate reservoirs. The water quality sensors measure a range of parameters, including temperature, dissolved oxygen, pH, and conductivity, among others. These parameters play a crucial role in determining water quality and ensuring it meets the required regulatory standards. The buoy based system is powered by solar panels that charge the onboard batteries, which powers the sensors, a transceiver, and a datalogger. The sensors transmit data to a Signal Conditioning and Data Acquisition (SCADA) system that provides real-time data to water treatment facility operators. The SCADA system is programmed to monitor the water quality data based on established criteria and send alerts to operators if there is any deviation from the acceptable standards. Operators can also adjust the sampling frequency remotely to investigate the water quality within the reservoirs more thoroughly.[13]

**Case 7.2: Acoustic Sensor Trial by Sydney Water for Leak Detection in Underground Water Network**
Sydney Water used 600 acoustic sensors to detect hidden leaks in its underground water network, saving 9,000 megalitres of water and cutting waste by $20 million. The sensors were placed on 13 kilometres of water mains in the Sydney CBD and designed to detect non-surfacing leaks up to 200 metres away by listening to the high-pitched frequencies from escaping water. Logger locations and leak alarms were made visible on a purpose-made web portal. The sensors were used in the early hours of the morning to prevent false positive results. The technology was successfully trialled over two years starting in 2019, discovering 160 hidden leaks and enhancing leak prevention efforts. The saved water is equivalent to 3,600 Olympic-size swimming pools, with accuracy of prediction reaching up to 95 percent. Sydney Water plans to integrate the sensors into its maintenance approach, reducing the need for reactive maintenance.[14]

**Case 7.3: United Utilities' Successful Trial of i2O's Advanced Pressure Management Solution**
United Utilities, a provider of water and wastewater services to 3 million households and 200,000 businesses across North West England, has successfully tested i2O's Advanced Pressure Management solution in the Lister Drive District Metered Area (DMA). The Lister Drive DMA has 1,009 domestic connections and

31 industrial connections, with a total mains length of 8,734m. United Utilities wanted to assess the leakage reduction capabilities of i2O's solution in this DMA, which was already deemed to have very good pressure control. When the Advanced Pressure Management solution was first installed, it was set to match the previous fixed outlet pressure. The solution is designed to continuously adapt to variations in demand and to react to changes in the network as they occur, enabling pressure at the critical point to be optimised for all network conditions. The solution was used to reduce pressure in the Lister Drive DMA by 1.0m intervals. Then the flow-modulation stage was initiated, with the system set to achieve a pressure of 20m at the specified minimum critical point. The implementation process was successfully completed, achieving and maintaining a target minimum pressure of 20m. The solution continued to operate reliably, reducing leakage steadily over the nine-month trial period. The successful trial resulted in significant benefits including 41.6 million litres of water saved per year, a 36% reduction in minimum night flow, leakage reduced by 114m$^3$/day, and a stable pressure regime achieved, resulting in fewer bursts and better customer service.[15]

## 7.1.2.2 SCADA systems

A water distribution network is made up of different operational components, including sensors, meters, pumps, and control valves. Components can be monitored or controlled onsite or from a central location. In the past, these operations were usually done using onsite instrument or control panels. In recent times, water utilities have transitioned to SCADA systems which measure, acquire information, and control over a distance. SCADA systems can process information and remotely operate and optimise systems and processes for a variety of uses. SCADA systems are used for accomplishing remote monitoring and control of water distribution facilities. The SCADA system's data acquisition function can be used for a variety of uses including monitoring storage tank levels, residual chlorine levels, pH levels, pressures, flows, pump status points (i.e. on/off), chemical feed station operation, and energy consumption. The SCADA system's control function can be used to optimise distribution system operations; for example, storage tank operating levels can be set based on real-time demand.[16,17]

#### Case 7.4: SA Water Enhancing its Reliability and Resilience with its new SCADA system
SA Water has recently upgraded its critical SCADA system to enhance the reliability and resilience of its water and sewerage infrastructure that caters to 1.7 million people. The improved SCADA system has enabled the utility to centralize and upgrade its critical control systems, resulting in the capacity to "monitor, control, upgrade and support our critical infrastructure on demand, with minimal service interruptions, and in any situation, from isolated issues to a state-wide power interruption or targeted cyber attack." The new system also utilizes a new type of programmable logic controller (PLC) that is specifically designed for water treatment plants. SA Water replaced its decentralized platform with a central, virtual solution located in a secure data center, thus achieving vast improvements in both physical and cyber security. The new system ensures faster operational support and disaster recovery with the ability to take operational control of multiple sites within minutes, quicker and easier cyber security techniques and patching, and immediate implementation of system rollbacks. The system's recovery time has reduced from four-to-five days to just hours. In cases where equipment is unable to be centrally operated, local workers at sites can re-take control using an emergency human-machine interface. SA Water's proj-

ect aligns with its strategic objective of getting the basics right every time, while also demonstrating the utility's commitment to leading the way with new technologies and fresh approaches, and working together with capable and committed teams to manage change. The new SCADA system is expected to provide benefits to SA Water for the next 15 years.[18]

### 7.1.2.3 GIS

A GIS serves as a repository of location information and asset details, based on a web map with layers corresponding to various systems that can be updated and shared in real-time with field crews. In water resources management, GIS is used to monitor water objects while checking the frequency and mapping of the quality of water sources. GIS can be applied in a variety of ways, including the following:

- *Asset management*: GIS enables water utilities to know in detail their assets, what their conditions are, what maintenance is required or the necessary budget, for example, it enables water utilities to determine whether the pipes that break the most often have a certain diameter or material
- *Disaster forecasting*: Flood reduction and drought monitoring programmes use GIS technology for forecasting. GIS determines the range of disaster events including magnitudes, frequencies, depth, and velocities
- *GIS in surface water and groundwater management*: Surface water risk management is determined by GIS, with data collected able to predict rainfall, determine the risk to aquatic habitat from surrounding areas, and assess pollution levels. GIS can be used to help measure the depth of groundwater as well as its quality. The groundwater source can be studied before drilling or developing a water source to reduce the risk of contamination[19,20,21]

**Case 7.5: Thames Water's ArcGIS emergency event management solution**

Thames Water, a water utility provider delivering 2.7 billion litres of drinking water per day to ten million customers, has implemented an ArcGIS Emergency Event Management solution to better respond to extreme weather events and unexpected pipe bursts that interrupt clean water supply. The solution includes three main components:

- *Event viewer*: an interactive dashboard that enables Thames Water to identify incidents, minimize their impact on customers, and provide a holistic view of the region to tailor responses.
- *Event manager*: a solution that provides a detailed view of incidents, improving communication between teams working on the ground and employees at head office, enabling Thames Water to make more rapid decisions during serious incidents, plan interventions more effectively, and allocate resources more appropriately.
- *Event reviewer*: a solution that enables Thames Water to learn from past incidents, assess the risks and actions taken, and use this insight to continuously improve its response to minimize or eliminate the impact of unplanned events on customers in the future.

Thames Water also added an ArcGIS mobile app to record changing water levels and new leaks in the field, which updates information in real-time on the Event Viewer dashboard. The new system enables

Thames Water to respond more quickly to incidents, improve customer service, and collaborate more effectively with stakeholders. The benefits include customer-focused responses during major incidents, faster restoration of water supplies, greater readiness for future emergencies, and improved collaboration with external organizations. The solution has helped Thames Water to meet new industry targets for restoring customers' clean water services within three hours.[22]

### 7.1.2.4 Software

Software is used to store, use, and report data. It can be used for modelling of infrastructure and environmental systems, decision-making, and risk management. Software is usually integrated with SCADA and/or GIS to manage water networks, control pressure, and monitor leakage. Software is also used for smart metering, billing and collections, hydrological modelling for water security, and cloud-based management and hosting options. For instance, online portals can be developed that provide customers with household consumption data at the yearly, monthly, daily, and even hourly level as well as historical consumption patterns, leak information, and water use comparisons with similar households, all in an interactive, web-based format. Customers who sign into these portals are often provided access to customisable leak detection alerts and notifications. Customers that set up automated leak alerts typically have them delivered to their smartphones when the system detects a leak on their property at a specified scale.[23,24]

> **Case 7.6: Dubai Electricity and Water Authority' High-Water Usage Alert service**
> Dubai Electricity and Water Authority (DEWA) has launched a High-Water Usage Alert service that has significantly reduced carbon emissions, by identifying and repairing water leaks, defects, and increased load. The service sends instant notifications to customers to conduct necessary maintenance to reduce water wastage, and since its launch three years ago, has reduced carbon emissions by 217,370 tons. DEWA provides this service as part of the Smart Living initiative to help customers detect any leakage in water connections after the meter. The service sends notifications to customers in case the smart meter detects any unusual rise in consumption to examine the internal connections and fix any leakage in water connections with the help of a technician. DEWA's smart infrastructure and advanced meters allow customers to monitor and manage their consumption proactively, which contributes to the sustainability of resources.[25]

### 7.1.3 Smart water grids and smart water meters

A smart water grid integrates ICT into the management of the water distribution system. Sensors, meters, digital controls, and analytic tools are used to automate, monitor, and control the transmission and distribution of water. Smart water grids aim to ensure water is efficiently delivered only when and where it is needed and that the water is of good quality. Smart water grids provide a wide range of benefits, including:

- *Real-time monitoring of asset condition and preventative maintenance*: With advanced sensors, data can be gathered on pipeline conditions and used to develop a risk-based model for pipe replacement projects. This enables utilities to better plan and schedule mains replacements and rehabilitation programmes
- *Real-time pressure and water quality monitoring*: Real-time sensor and meter data allow water utilities to quickly detect leaks to minimise water losses as well as detect stress in pipes early to mitigate the risk of pipe bursts. Water utilities can also use this technology to continuously monitor water quality in the distribution pipelines, providing early warning of potential contamination
- *Real-time water consumption information to help customers conserve water*: Smart water grids and smart water meter technology enables water utilities to provide customers with real-time feedback on water, as well as energy, usage. This helps customers make informed choices with regards to water consumption. Also, usage data from smart meters enable more accurate demand prediction for optimising water pumping schedules and the volume of water required to treat and pump[26,27]

**Case 7.7: SA Water's Smart Water Network**
SA Water has successfully implemented smart network technology that is detecting and repairing water main breaks and leaks in Adelaide's central business district. The $4 million smart network trial uses over 300 acoustic detection sensors that can monitor and identify acoustic sounds indicating potential cracks or leaks in pipes. These sensors cover an average range of 100 metres and monitor around 50% of the city's water main network, with a focus on cast iron pipes in areas where the potential customer impact of a break is greater. In addition to acoustic sensors, the smart network comprises various IoT-enabled sensors such as smart meters, mass pressure sensors, water quality sensors, mass flow meters, and pressure transient/hydrophone sensors. Since its inception, the smart network has successfully detected over 50% of all water main breaks and leaks in the Adelaide CBD. Out of 76 faults, 40 were detected and repaired proactively, compared to 36 that were detected and repaired reactively. Reactive incidents were sudden failures that did not offer any detectable signs, but SA Water's smart network is helping to improve their understanding of acoustic patterns to identify circumferential or longitudinal cracks in pipes. Following the trial's success, SA Water has expanded its smart network to six further locations across South Australia, including sewer and odour monitoring. The success of SA Water's smart network technology highlights its potential to reduce costs and community disruption while also improving network management practices in the water industry.[28]

### 7.1.4 Artificial intelligence and machine learning

Artificial intelligence (AI) is intelligence exhibited by machines or computers, allowing them to perform tasks such as understanding, learning, reasoning, planning, and more. In its current application, AI systems can rationally solve complex problems, predict outcomes, and act in real-world situations to achieve goals. The spectrum of AI is expanding and includes:
- Automated intelligence systems that take repeated, labour-intensive tasks requiring intelligence, and automatically completes them

- Assisted intelligence systems review and reveal patterns in historical data and help people perform tasks more quickly and better use the information collected
- Augmented intelligence systems that use AI to help people understand and predict an uncertain future
- Autonomous intelligence systems that automate decision-making without human intervention

In the context of water resources management, a wide variety of IoT sensors and other data-driven technologies can continuously collect data on different phases in the water supply and demand. As such, AI can be used in a variety of contexts as summarised in Table 7.2.[29,30,31]

**Table 7.2**: Contexts that AI Can be Used in Water Resources Management.

| Context | Example |
| --- | --- |
| Water supply | – Water supply monitoring and management<br>– Water quality simulation and data alerts<br>– Asset management on critical water and wastewater expenditures |
| Catchment control | – Harmful algal blooms detection and monitoring<br>– Streamflow forecasting<br>– Automated flood-centred infrastructure |
| Water efficiency | – Residential water use monitoring and management<br>– Optimisation of industrial water use<br>– Predictive maintenance of water plants<br>– An early-warning system for water infrastructure<br>– Detect underground leaks in potable water supply systems<br>– Smart meters in homes |
| Adequate sanitation | – Drones and AI for real-time monitoring of river quality<br>– Ensuring adequate sanitation of water reserves<br>– Real-time monitoring and management of household water supply |
| Drought planning | – Drought prediction<br>– Simulations for drought planning<br>– Drought-impact assessments |

### 7.1.4.1 Machine learning

Machine learning (ML) is a subset of AI that helps derive meaning from data generated by people, devices, and smart systems etc. Increasingly, the volume of data collected is surpassing the ability of humans to make sense of it and use it efficiently. ML uses this data to create predictions or answer questions. Specifically, a predictive model is trained using data to create predictions or answer questions, with the more data gathered, the more the model can be refined, and new predictive models developed.[32]

**Case 7.8: Yorkshire Water Trialling Artificial Intelligence to Improve Leak Detection**
Yorkshire Water has collaborated with Siemens and Artesia Consulting to deploy AI in a trial aimed at detecting leaks more effectively. Currently, nearly 40,000 acoustic loggers are deployed across Yorkshire Water's clean water distribution network to listen for evidence of leaks each morning. However, sometimes these loggers raise false positives due to background noise. The project aims to reduce the number of false alarms by up to 60 percent by using trained analytics packages that can distinguish between a leak and background noise through machine learning techniques and analysis of audio recordings. The system can simultaneously identify more true leaks, which will allow Yorkshire Water to reduce leakage by targeting activities more effectively. Siemens data science experts from multiple sectors collaborated with Yorkshire Water's deep knowledge of water networks to train a powerful AI that can analyse complex acoustic data. Artesia Consulting improved the accuracy and sensitivity of the technology through data science and industry expertise.[33]

## 7.2 Managing customers of the future

Water utilities need to move away from viewing customers as recipients of services and instead view them as active participants in the delivery of those services.

Customers and communities also have knowledge, skills, and creativity that can solve problems and help find ways to innovate. At the same time, there are growing customer expectations of the level of service delivered by water utilities.[34,35]

### 7.2.1 Customer participation

Ofwat defines customer participation as *"the active involvement of customers in the design, production, delivery, consumption, disposal and enjoyment of water, water services and the water environment in the home, at work and in the community"*. There are different levels of customer engagement and involvement that water utilities can incorporate in their strategies to enhance customer participation, including:

- *Level 1: Listening and understanding*: Understanding what is important about water in the lives of different customer groups
- *Level 2: Listening and acting*: Listening to different customer groups and acting on what is heard to achieve business objectives
- *Level 3: Engaging and involving*: Involving customers or their representatives by making it easy for them to propose specific ideas or solutions to achieve change
- *Level 4: Customer participation*: Increasing active customer participation to bring these ideas to life

There is a range of benefits that customers can receive through participation including influencing the future, protecting lifestyles, improving local environments, improving customer service, saving money, saving water, avoiding the risk of flooding, avoiding the risk of sewer flooding, and feeling in control.[36]

### 7.2.1.1 Strategic framework to increase customer participation

Ofwat has proposed a strategic framework (Table 7.3) that water utilities can follow when aiming to increase customer participation. The framework allows a water utility to achieve a variety of ambitions, including:

- Co-imaging the future with their customers
- Co-creating the future with their customers
- Engaging customers to adopt actions or behaviours at scale to achieve real change
- Engaging citizens to own improvements to water resilience in their communities
- Giving customers more control over water in their homes
- Giving customers more control over their service experience[37]

**Table 7.3:** Strategic Framework to Increase Customer Participation.

| Area | Aim | Benefits |
|---|---|---|
| *Futures*: Customer participation in the sector's future | Enhancing customer participation to improve the current and future sustainability of water in the lives of customers | – Increased customer support for plans<br>– Improvements to customer satisfaction and customer trust<br>– Innovative ideas from customers to help achieve sector goals<br>– Active engagement to create a resilient future for water |
| *Action*: Increasing customer action to improve resilience | Increasing customer behaviour change actions, including saving water | – Increases the opportunity for financial benefits<br>– Reduces costs<br>– Improves sector resilience |
| *Community*: Increasing community ownership and participation | People acting together in local areas can make improvements to their local water environment | – A feeling of shared ownership in local communities<br>– Increased understanding of the importance of water<br>– Peer group persuasion |
| *Experience*: Increasing participation in the customer experience | Increasing customer control of water in their home or of the customer service experience | – Customer satisfaction increasing following contact<br>– Reduced repeat calls on customer service issues<br>– Product and service improvement ideas from customers |

### 7.2.1.2 Strategies and tools to increase customer participation

A variety of strategies and tools can be implemented by water utilities to increase customer participation, including the following:

- *Personalisation*: Customer participation strategies need to make the customer feel special which promotes loyalty, which in turn makes the customer become an ambassador of the brand
- *Exclusivity*: Customer participation can reward loyal customers with access, information, and exclusive offers. Making a customer feel like a VIP enhances the connection between the customer and the brand, which increases the likelihood of the customer promoting the brand inside their social environment online and offline
- *Smartphone apps*: Smartphone apps provide customers with new ways of managing, calculating, communicating, and evaluating environmental information, potentially empowering customers to play an important role in the promotion of sustainable consumption
- *Gamification*: Gamification is the use of game designs in non-game contexts. Gamification encourages customers to adopt the behaviour associated with the game. Specifically, gamification can guide and motivate customers to change their behaviours and achieve meaningful long-term objectives. Customers can also share their game results with friends, increasing the social network influence of the brand[38,39,40]

**Case 7.9: Personalisation: Austin's Green Business Leaders saving water**
The City of Austin runs the Austin Green Business Leaders programme which offers official recognition for businesses that take actions to protect the local environment while at the same time saving money and attracting new customers. Regarding water, Austin Green Business Leaders can participate in the 3C Business Challenge. To receive the 3C Business Challenge Certificate, businesses must:
- Sign and submit a completed form and checklist
- Identify and fix any leaks and adjust existing water-using equipment so that they are operating efficiently and without waste
- Make sure plumbing fixtures and water using equipment comply with existing codes and ordinances
- Evaluate the replacement or retrofit of equipment to more efficient models using Austin Water rebate opportunities or successfully participate in the utility's audit rebate programme

Overall, businesses that excel in the Austin Green Business Leaders programme are formally recognised as Silver, Gold, or Platinum Green Business leaders, have their business featured on the City's website and can place the programme logo on the storefront window and company website, and join the network of Austin's top sustainable companies, enabling them to make new connections, share best practices, and learn from peers at members-only events.[41]

**Case 7.10: Exclusivity: Ventura Water's Capture Conservation contest**
Over the summer period of 2016, Ventura Water in California ran its Capture Conservation contest which aimed to highlight the community's response to the call to water conservation. As part of Ventura Water's annual summer awareness campaign, residents were invited to share photos of how the drought has inspired them to save water. Contestants took a photo that captured their water conservation story and posted the photo to Instagram, Facebook, or Twitter with the hashtag #keepsavingventura. The winners went above and beyond the call of duty to minimise their water footprint. Contest winners were awarded the following prizes:

- Five Deluxe car washes compliments of Final Details
- Four Golf N' Stuff tickets
- 40 percent off coupon to Patagonia
- Two tickets to Santa Cruz Island
- $100 gift card to Green Thumb Nursery[42]

### Case 7.11: Smartphone apps: Watercare's smart phone app

Watercare, a water and wastewater service provider in Auckland, New Zealand, offers a smart meter app that helps customers manage their water usage more efficiently. The smart meter app provides a convenient way for customers to monitor their water consumption, detect potential leaks, and make informed decisions about their water use. Key features of the Watercare Smart Meter app include:

- Usage monitoring: Customers can track their daily water consumption and compare it with previous periods, helping them identify trends and potential areas for improvement. The app displays usage data in an easy-to-understand format, making it simple for users to analyze their water consumption patterns.
- Leak detection: The smart meter app alerts customers if it detects continuous water flow, which could indicate a potential leak. By identifying leaks early, customers can address issues promptly and minimize water waste.
- Consumption goals: The app allows users to set water consumption goals and track their progress, encouraging more efficient water usage and promoting water conservation.
- Notifications: Customers can opt to receive notifications about their water usage, helping them stay informed and manage their consumption more effectively.
- Bill estimation: The smart meter app provides an estimated bill based on the customer's water usage, enabling them to anticipate and budget for their upcoming water charges.[43]

### Case 7.12: City of Davis' Gamification in Water Conservation

The City of Davis, a California water utility, has embraced an innovative strategy to engage customers and promote water conservation by using interactive online video games. These games aim to educate players about the importance of a resilient water supply and encourage sustainable water use practices. Collaborating with local college students and an outreach consultant, the City of Davis developed a series of interactive online games designed to teach players about preserving the natural environment, reducing water waste, and preventing pollution. This approach, known as "gamification," incorporates game design elements into non-game contexts, such as educational settings, to enhance motivation and focus on the task at hand. The games serve as an additional outreach tool, engaging customers of various ages. For older players, the games challenge preconceived notions and present alternative approaches in a fun and non-confrontational manner.

The environmental video games were designed to be compatible with the Chromebook platform used by students in local schools. Players can acquire tips on water conservation, learn about recyclable and compostable products, and discover what should not be discarded into storm drains by matching images or identifying water leaks on a city map. The outreach consultant developed the water conservation games, while city staff members designed graphics for the stormwater and recycling games to complement the code written by the college students. The city hosts these games on its website and promotes them through social media, local events, and newsletters.[44]

### 7.2.1.3 Social media strategies to increase customer participation

Water utilities are utilising a range of social media channels, including Twitter, Facebook, Instagram, and YouTube among others, to engage with customers, build brand visibility and popularity, inform customers of products, and offer customer services. Social media strategies that aim to increase customer participation can include the following:

1.  *Build consumer connections*: Most often a water utility's social media practice revolves around outage notifications and communication, with water utilities using social media channels to proactively inform customers about a planned outage and quickly respond to a large base of customers regarding an outage, breakdown or disruption due to natural calamities. Water utilities can integrate their social media communication with their CRM function with social media communications going out in a few seconds of an outage, followed by calls, emails, and physical customer care visits within hours of the outage

2.  *Create customer awareness*: Water utilities can use social media to educate their customers on topics such as water conservation, advantages of smart meters, and industry trends. Social media can also be utilised to generate user-specific awareness regarding changes in pricing or billing. Smart meter customer awareness revolves around increasing awareness on issues such as 'what are the benefits for customers', 'how can the consumer manage consumption', 'how can the consumer be impacted during installation'

3.  *Create brand awareness*: Customers are increasingly using social media to build or destroy the reputation of their service providers. Social media provides water utilities with the opportunity to manage brand perception and map customer sentiments towards the brand. Brands can also use their social media presence to manage customer satisfaction effectively

4.  *Offer water advice and tips*: Social media can be used to educate customers about water conservation and water efficiency technologies. Water utilities can use social media to reach out to customers rather than adopting the expensive traditional ways of creating and managing audio-visual campaigns. This mode can be used to promote web-based tools to help consumers analyse their monthly usage and work towards a lower bill. Many water utilities are pursuing two-way discussions with customers on ways to save water, as well as energy, and the importance of water conservation. Real-time communication with customers across multiple platforms tends to increase engagement and allow 'virtual' conversations with the customer[45]

**Case 7.13: Summary of social media strategies to engage customers**
A few water utilities have initiated a range of social media strategies on Twitter to engage customers, examples of which are provided in Table 7.4.[46]

**Table 7.4:** Examples of Social Media Strategies on Twitter to Engage Customers.

| Water Utility | Social Media Strategy |
|---|---|
| San Antonio Water System | San Antonio Water System's Twitter feed includes educational information, for example, what to flush/what not to flush/where to take the things you should not flush, community cheerleading, for example, 'go, go high schoolers', and talk on food, in particular on good tacos |
| Northumbrian Water | Northumbrian Water's Twitter feed includes shots of the region's waterways and protection efforts, strangest things to find in pipes, along with a mix of serious and humorous tweets |
| San Jose Water | San Jose Water tweets almost exclusively about customer desires, from outages to meet-and-greets. They sometimes provide a recipe that can be made with San Jose water |
| Yarra Valley Water | Yarra Valley Water informs its customers daily on issues along with tweets to encourage the use of reusable bottles and to 'Be Smart. Choose Tap' |
| Southern Water | Southern Water provides educational polls, heartfelt videos, and detailed snapshots of daily life inside the utility |

With regards to measuring customer participation across social media, there are three main metrics available that can be followed up:

- *Commitment metrics*: Commitment metrics are based on the percentage of customers committed, which is determined by basics such as site traffic, fans, followers, likes, and shares etc.
- *Customer metrics*: Customer metrics can be used to focus on enhancing customers' loyalty, for example, net promoter score is the percentage of customers rating their likelihood of recommending a company, a product, or a service to a friend or colleague
- *Financial impact*: Financial impact involves identifying customer profiles and then conducting targeted campaigns such as contests and promotions to determine a return on investment[47]

### 7.2.2 Enhancing customer experiences across the water distribution network

There are a variety of strategies water utilities can implement to enhance customer experiences across the water distribution network, including the following summarised in Table 7.5.[48,49]

**Table 7.5:** Strategies to Enhance Customer Experiences.

| Strategy | Description | Examples |
|---|---|---|
| Analytics | Predictive analytics and modelling can pre-empt issues and encourage proactive action by customers | – Postcode-driven early warning messages can inform customers of planned works that will reduce water pressure<br>– Integrating weather and temperature data can help customers take preventative action to avoid frozen pipes or anticipate the risk of flooding |
| Automation | Automation can deliver service improvements with new technologies such as robots and sensors able to detect problems in the water network before they affect customers | – Robots can detect leaks in the distribution system<br>– Sensors can detect water quality issues in reservoirs and water pipelines |
| Influencing behaviour change | Smartphone apps and digital devices can deliver timely and targeted prompts to help people to keep track of the water they are using | – A higher than normal bill could trigger water-saving advice along with examples of how much water can be saved on the next bill via Internet portals or app channels |
| Community engagement | Customer engagement can extend beyond functional customer service contact and problem resolution. Utilities can run schemes that help local communities | – Online community platforms make it easier for customers to engage with utilities on issues<br>– AI can deliver bespoke flooding information to customers, making sure they know how to respond when problems occur<br>– Interactive maps can be used to enable customers to report leaks |
| Self-service | Self-service allows customers to forgo call centres. This means water utilities are more accessible and responsive to customers 24/7. Developments in conversational technology including chatbots, voice interfaces, and conversational search are helping customers find the information they need more easily | – Apps that allow customers to make payments, submit meter readings, and view payment history<br>– Report leaks |

**Case 7.14: SMS Alert Service for Monitoring Water Levels and Heavy Rain in Singapore**

Individuals interested in monitoring water levels in specific canals or drains or receiving notifications about heavy rain forecasts in Singapore can register for an SMS alert service. Subscribers can opt for both water level and heavy rain alerts or choose either one. Each mobile number is allowed a single subscription for a selected location for water level alerts. There are two types of notifications available: Heavy Rain Notifications and High Water Level Notifications. Regarding Heavy Rain Notifications, upon subscription, participants will receive SMS alerts from the Meteorological Service Singapore (National Environment Agency (NEA)) when heavy rain is anticipated to impact Singapore. The alert will include details about the expected location of the heavy rain and tidal change data if high tide coincides with heavy rainfall. Subscribers will receive another SMS when heavy rain is no longer predicted or has stopped. Tide Times can be viewed online. Regarding High Water Level Notifications, as part of its extensive drainage management and control efforts, Public Utilities Board (PUB) has installed 210 sensors in crucial canals and drains to monitor water levels. Subscribers will receive incremental SMS alerts if the water level in the selected canal exceeds 50%, 75%, 90%, and 100% of the canal's depth. Additional SMS alerts will be sent to inform subscribers when the water level gradually drops below 50%. PUB Subscribe to SMS Alerts.[50]

**Case 7.15: Automation: Singapore's autonomous drones monitoring reservoirs**

Singapore's PUB has deployed autonomous drones to manage aquatic plant growth and monitor reservoir activities. The Beyond Visual Line of Sight (BVLOS) drones will begin monitoring six reservoirs across Singapore, with each drone equipped with remote sensing systems and a camera for near real-team video analytics. Starting with one drone at MacRitchie and another at Marina reservoir, PUB will progressively deploy one drone each to the four other reservoirs (Serangoon, Kranji, Lower Seletar, and Lower Peirce) later this year. The drones at MacRitchie and Marina are housed in an automated pod and are capable of taking off and landing autonomously. The flights are pre-programmed and monitored remotely by an operator. The drones will conduct daily patrols to monitor water quality, detect excessive growth of aquatic plants and algal blooms, and collect data on water activities, such as fishing and paddling, in and around the reservoirs' edge. Specifically:

- *Monitoring water quality*: Using the drone's water quality remote sensing system, the drone will gather data on turbidity and algae concentration, which provides a good correlation to actual water quality. If necessary, PUB officers will collect water samples on-site for further analysis
- *Monitoring aquatic plant growth*: A video analytics algorithm is used to identify aquatic plant overgrowth in the reservoir via the drone camera's live video feed. The algorithm can differentiate between plants above the surface and plants that are submerged.
- *Monitoring water activities*: The drone's live video feed will monitor water activities, with a video analytics algorithm flagging potential concerns, such as anglers fishing in non-designated areas or overcrowding of vessels in a particular area. If any concerns arise, a message will be sent to a dedicated Telegram channel that officers can access via their mobile phones, allowing them to respond to issues quickly

Overall, 7,200 hours of staff time is normally used per annum to undertake these tasks. By utilising the BVLOS drones, around 5,000 hours of staff time will be saved, allowing the utility to redirect its staff to other works.[51]

**Case 7.16: San Diego's Waste No Water Smart App**

The City of San Diego's "Waste No Water" Smart App is available for both iPhone and Android users and is designed to help individuals play their part in reducing water waste. The app encourages responsible water use by enabling users to report instances of water waste in their community. By simply taking a photo of the violation and entering the address, users can submit their complaint to the city's Conservation Department. A complaint file number is generated, and the issue will be addressed by the relevant staff. The app also helps educate the public on the restrictions in place and what constitutes wasteful water use. Users can track the complaint they submitted, ensuring that the issue has been addressed and that water conservation measures are being taken seriously. In addition to its reporting function, the "Waste No Water" Smart App also provides users with helpful tips and advice on how they can conserve water in their daily lives. This includes information on water-saving devices, best practices for efficient irrigation and ways to reduce indoor water usage. The app also provides users with up-to-date information on the city's water restrictions and conservation efforts, allowing them to stay informed and be part of the solution. By encouraging people to be proactive in reducing water waste, the "Waste No Water" Smart App is helping to create a more sustainable future for everyone. One of the key benefits of this app is that it allows individuals to take an active role in conserving water. By reporting instances of water waste, users are contributing to a more sustainable future and helping to ensure that everyone has access to clean and safe water. The app also helps to create a sense of community, with users working together to conserve water and educate each other on best practices. The app is also cost-effective, as it allows the city to address water waste complaints quickly and efficiently. The app's GPS feature helps the city's Conservation Department locate the issue and respond promptly, reducing the amount of time and resources needed to resolve the problem.[52]

**Case 7.17: Anglian Water's Customers Cooperating in Reporting Water Leaks**

Anglian Water operates in the driest region of the UK, where it relies on the cooperation of its customers to conserve water by reporting all water leaks, regardless of their size. The company urges the public to notify them immediately if a leak is noticed beneath a pavement or road, as this will enable swift repairs.

Anglian Water has achieved considerable success in addressing water leaks, managing to supply the same volume of water daily as it did two decades ago despite a 20% increase in customers. The company acknowledges that this achievement is only possible with the help of its customers. To report a leak, customers can use either an online map or call Anglian Water's 24/7 leakline number for prompt action. The more information provided about the leak, the quicker it can be located and repaired. The following details should be included when reporting a leak to Anglian Water:

–   The exact location of the leak, indicated by dropping a marker on a Google map.
–   Contact details of the person reporting the leak, in case Anglian Water needs to follow up.

For leaks on a customer's supply pipe, Anglian Water provides guidance on who is responsible for the repairs and offers a simple seven-step process for detecting leaks if a water meter is installed. Furthermore, the company offers advice for customers experiencing a burst pipe in their home.[53]

**Case 7.18: Self service: United Utilities' Smartphone App Enhancing Customer Experience**

United Utilities, a leading water and wastewater company in the United Kingdom, has developed a smartphone app designed to improve customer experience and streamline access to essential services. This user-friendly app allows customers to efficiently manage their water accounts, making it easier than ever to monitor water usage, report issues, and make payments. The key Features of the United Utilities App are:

- *Account management*: Customers can easily view their account details, check their payment history, and access past statements. The app simplifies the process of understanding and managing water usage, which can lead to more informed decisions and better conservation efforts.
- *Meter reading submission*: The app facilitates the submission of meter readings, ensuring accurate billing and helping customers track their water consumption. With a user-friendly interface, submitting readings is quick and straightforward.
- *Bill payment*: Users can pay their water bills securely and conveniently through the app. The payment process is simple, and customers can manage their payment preferences with ease.
- *Leak reporting*: The United Utilities app includes a feature for reporting leaks, allowing customers to contribute to the efficient maintenance of the water infrastructure. The app's integrated map enables users to pinpoint the exact location of the leak and even upload photos for better issue resolution.
- *Flashlight and camera function*: To assist customers in reading their water meters, the app includes a flashlight function and camera integration. These features make it easier to submit accurate meter readings, even in low-light conditions.[54]

# Notes

1  Ofwat, "Tapped in – from Passive Customer to Active Participant Report," (2017), https://www.ofwat.gov.uk/publication/tapped-in-from-passive-customer-to-active-participant/.

2  ADB, "Public-Private Partnerships and Smart Technologies for Water Sector Development," (2013), http://events.development.asia/system/files/materials/2018/07/201807-public-private-partnerships-and-smart-technologies-water-sector-development-summary.pdf.

3  IWRA and K-Water, "Smart Water Management Case Study Report," (2018), https://www.iwra.org/swmreport/.

4  ADB, "Public-Private Partnerships and Smart Technologies for Water Sector Development".

5  IWRA and K-Water, "Smart Water Management Case Study Report".

6  Ibid.

7  Kazeem B. Adedeji et al., "Pressure Management Strategies for Water Loss Reduction in Large-Scale Water Piping Networks: A Review" (paper presented at the Advances in Hydroinformatics, Singapore, 2018// 2018).

8  WHO, "Leakage Management and Control – a Best Practice Training Manual," https://www.who.int/docstore/water_sanitation_health/leakage/begin.html#Contents.

9  S. Geetha and S. Gouthami, "Internet of Things Enabled Real Time Water Quality Monitoring System," *Smart Water* 2, no. 1 (2017).

10  A. K. Mamun et al., "Smart Water Quality Monitoring System Design and Kpis Analysis: Case Sites of Fiji Surface Water," *Sustainability* 11, no. 24 (2019).

11  Wesley Schultz, Shahram Javey, and Alla Sorokina, "Smart Water Meters and Data Analytics Decrease Wasted Water Due to Leaks," *Journal – AWWA* 110, no. 11 (2018).

12  Nourhan Samir et al., "Pressure Control for Minimizing Leakage in Water Distribution Systems," *Alexandria Engineering Journal* 56, no. 4 (2017).

**13** Xylem, "Nyc Monitors Source Waters with Ysi Buoys," (2021), https://www.ysi.com/file%20library/documents/application%20notes/a507-monitoring-source-water-before-it-reaches-treatment-plants-in-new-york-city.pdf.
**14** iTnews, "Sydney Water Trials Acoustic Sensors for Leaks, Saves 9000 Megalitres," https://www.itnews.com.au/news/sydney-water-trials-acoustic-sensors-for-leaks-saves-9000-megalitres-580664.
**15** i20, "United Utilities," https://en.i2owater.com/clients/united-utilities/.
**16** G.J. Kirmeyer and AWWA Research Foundation, *Guidance Manual for Maintaining Distribution System Water Quality* (AWWA Research Foundation and American Water Works Association, 2000).
**17** Kentucky Water Resources Research Institute, "Water Distribution System Toolkit," http://www.uky.edu/WDST/SCADA.html.
**18** ITNews, "Sa Water Boosts Resiliency of Clean Water, Sewerage Services," https://www.itnews.com.au/news/sa-water-boosts-resiliency-of-clean-water-sewerage-services-538538.
**19** ADB, "Public-Private Partnerships and Smart Technologies for Water Sector Development".
**20** Umwelt und Informationstechnologie Zentrum, "Gis in Water Resource Monitoring," http://uizentrum.de/en/gis-in-water-resource-monitoring-2/
**21** Software Advice, "How to Optimize Utility Asset Management with Gis," https://www.softwareadvice.com/resources/optimize-utility-asset-management-with-gis/.
**22** Esri UK, "Thames Water," (2019), https://resource.esriuk.com/wp-content/uploads/Esri-UK-Thames-Water-14.7.20.pdf.
**23** Schultz, Javey, and Sorokina, "Smart Water Meters and Data Analytics Decrease Wasted Water Due to Leaks."
**24** ADB, "Public-Private Partnerships and Smart Technologies for Water Sector Development".
**25** Government of Dubai, "Dewa's High-Water Usage Alert Service Helps to Reduce 217,370 Tons of Carbon Emissions," https://www.mediaoffice.ae/en/news/2022/August/16-08/DEWAs-High-Water-Usage-Alert-service-helps.
**26** Singapore Public Utilities Board, "Managing the Water Distribution Network with a Smart Water Grid," *Smart Water* 1, no. 1 (2016).
**27** A. Cominola et al., "Data Mining to Uncover Heterogeneous Water Use Behaviors from Smart Meter Data," *Water Resources Research* 55, no. 11 (2019).
**28** SA Water, "Continued Sensor Success for Sa Water's Smart Network," https://www.sawater.com.au/news/continued-sensor-success-for-sa-waters-smart-network.
**29** PwC, "Fourth Industrial Revolution for the Earth Harnessing Artificial Intelligence for the Earth," (2018), https://www.pwc.com/gx/en/sustainability/assets/ai-for-the-earth-jan-2018.pdf.
**30** IWA, "Ai Basics for Advanced Water Wise Utilities – Part 1," https://iwa-network.org/ai-basics-for-advanced-water-wise-utilities-part-1/.
**31** Silo.AI, "How Artificial Intelligence Is Transforming the Water Sector: Case Ramboll," https://silo.ai/how-artificial-intelligence-is-transforming-the-water-sector-case-ramboll/.
**32** IWA, "Ai Basics for Advanced Water Wise Utilities – Part 1".
**33** Yorkshire Water, "Promising Results for Ai Leak Detection Trial," https://www.yorkshirewater.com/news-media/news-articles/2021/yorkshire-water-sees-promising-results-in-ai-leak-detection-trial/.
**34** Ofwat, "Tapped in – from Passive Customer to Active Participant Report".
**35** C. D. Beal and J. Flynn, "Toward the Digital Water Age: Survey and Case Studies of Australian Water Utility Smart-Metering Programs," *Utilities Policy* 32 (2015).
**36** Ofwat, "Tapped in – from Passive Customer to Active Participant Report".
**37** Ibid.
**38** Olli Tyrväinen, Heikki Karjaluoto, and Hannu Saarijärvi, "Personalization and Hedonic Motivation in Creating Customer Experiences and Loyalty in Omnichannel Retail," *Journal of Retailing and Consumer Services* 57 (2020).

**39** Andreas B. Eisingerich et al., "Hook Vs. Hope: How to Enhance Customer Engagement through Gamification," *International Journal of Research in Marketing* 36, no. 2 (2019).

**40** Christian Fuentes and Niklas Sörum, "Agencing Ethical Consumers: Smartphone Apps and the Socio-Material Reconfiguration of Everyday Life," *Consumption Markets & Culture* 22, no. 2 (2019).

**41** City of Austin, "Austin Green Business Leaders," https://austintexas.gov/department/austin-green-business-leaders.

**42** Ventura Water, (2016), https://www.cityofventura.ca.gov/DocumentCenter/View/8309/Pipeline_nov16_print?>bidId=

**43** Watercare, "Tap into Our App," https://www.watercare.co.nz/Help-and-advice/Smart-meter-app.

**44** AWWA, "Gamification Boosts Utility's Customer Engagement, Water Efficiency Success," https://www.awwa.org/AWWA-Articles/gamification-boosts-utilitys-customer-engagement-water-efficiency-success.

**45** WNS, "The Social Media Manual for the Utility Industry: Guidelines, Best-Practices and-Outsourcing-Strategies," https://www.wns.com/insights/articles/articledetail/71/the-social-media-manual-for-the-utility-industry-guidelines-best-practices-and-outsourcing-strategies.

**46** Oracle, "New Additions! The Best Water Utility Social Media Accounts You Should Be Following Right Now," https://blogs.oracle.com/utilities/the-best-water-utility-social-media-accounts-you-should-be-following-right-now.

**47** A. Moreno-Munoz et al., "Mobile Social Media for Smart Grids Customer Engagement: Emerging Trends and Challenges," *Renewable and Sustainable Energy Reviews* 53 (2016).

**48** Ecoconsultancy, "How Can Water Companies Use Digital to Improve Customer Experience?," https://econsultancy.com/digital-customer-experience-water-ofwat/.

**49** West Monroe, "3 Enablers to Water Utility Customer Centricity," https://blog.westmonroepartners.com/does-your-water-utility-have-a-customer-centric-workforce/.

**50** PUB, "Subscribe to Sms Alerts," https://www.pub.gov.sg/drainage/floodmanagement/subscribesms.

**51** Robert C. Brears, "Autonomous Drones Monitoring Singapore's Reservoirs," Mark and Focus, https://medium.com/mark-and-focus/autonomous-drones-monitoring-singapores-reservoirs-4abd3cddb622.

**52** "Tackling the Water Crisis: San Diego's 'Waste No Water' Smartphone App Leads the Way in Conservation," Mark and Focus, https://medium.com/mark-and-focus/the-smart-app-to-waste-no-water-4375684b13bc.

**53** Anglian Water, "Report a Leak," https://www.anglianwater.co.uk/services/water-supply/leakage/report-a-leak/#:~:text=Such%20is%20our%20success%20in%20tackling%20leaks%2C%20we%27re,years%20ago%20and%20that%27s%20with%2020%25%20more%20customers.

**54** United Utilities, "App," https://www.unitedutilities.com/app/.

# References

ADB. "Public-Private Partnerships and Smart Technologies for Water Sector Development". (2013). http://events.development.asia/system/files/materials/2018/07/201807-public-private-partnerships-and-smart-technologies-water-sector-development-summary.pdf.

Adedeji, Kazeem B., Yskandar Hamam, Bolanle T. Abe, and Adnan M. Abu-Mahfouz. "Pressure Management Strategies for Water Loss Reduction in Large-Scale Water Piping Networks: A Review". Paper presented at the Advances in Hydroinformatics, Singapore, 2018// 2018.

Beal, C. D., and J. Flynn. "Toward the Digital Water Age: Survey and Case Studies of Australian Water Utility Smart-Metering Programs". *Utilities Policy* 32 (2015/03/01/ 2015): 29–37.

Brears, R.C. "Hamburg'S Online, Real-Time Heavy Rain Map". Mark and Focus, https://medium.com/mark-and-focus/hamburg-s-online-real-time-heavy-rain-map-407e64777cc9.

_____. "The Rise of the Machines (in Managing Water)". Mark and Focus, https://medium.com/mark-and-focus/the-rise-of-the-machines-in-managing-water-96e8c0426178.

_____. "Singapore's Robotic Swans Testing Water Quality". Mark and Focus, https://medium.com/mark-and-focus/singapores-robotic-swans-testing-water-quality-30a97b666679.

_____. The 'Waste No Water' Smart App". Mark and Focus, https://medium.com/mark-and-focus/the-waste-no-water-smart-app-a54d0bef738f.

City of Austin. "Austin Green Business Leaders". https://austintexas.gov/department/austin-green-business-leaders.

Cominola, A., K. Nguyen, M. Giuliani, R. A. Stewart, H. R. Maier, and A. Castelletti. "Data Mining to Uncover Heterogeneous Water Use Behaviors from Smart Meter Data". _Water Resources Research_ 55, no. 11 (2019/11/01 2019): 9315–33.

Cominola, Andrea, Rohan Nanda, Matteo Giuliani, Dario Piga, Andrea Castelletti, Andrea-Emilio Rizzoli, Alexandros Maziotis, Paola Garrone, and Julien Harou. "The Smarth2o Project: A Platform Supporting Residential Water Management through Smart Meters and Data Intensive Modeling". In _American Geophysical Union_ 2014 _Fall Meeting_, 2014.

DC Water. "High Usage Alerts". https://www.dcwater.com/high-usage-alerts.

Ecoconsultancy. "How Can Water Companies Use Digital to Improve Customer Experience?" https://econsultancy.com/digital-customer-experience-water-ofwat/.

Eisingerich, Andreas B., André Marchand, Martin P. Fritze, and Lin Dong. "Hook Vs. Hope: How to Enhance Customer Engagement through Gamification". _International Journal of Research in Marketing_ 36, no. 2 (2019/06/01 2019): 200–15.

Esri. "Transforming Business Processes Enterprise-Wide". https://resource.esriuk.com/esri-resources/thames-water/.

Essex and Suffolk Water. "Leaks in Your Area". https://eswcommunityportal.co.uk/Leaks

Fuentes, Christian, and Niklas Sörum. "Agencing Ethical Consumers: Smartphone Apps and the Socio-Material Reconfiguration of Everyday Life". _Consumption Markets & Culture_ 22, no. 2 (2019/03/04 2019): 131–56.

Geetha, S., and S. Gouthami. "Internet of Things Enabled Real Time Water Quality Monitoring System". _Smart Water_ 2, no. 1 (2017/07/27 2017): 1.

IWA. "Ai Basics for Advanced Water Wise Utilities – Part 1". https://iwa-network.org/ai-basics-for-advanced-water-wise-utilities-part-1/.

IWRA and K-Water. "Smart Water Management Case Study Report". (2018). https://www.iwra.org/swmreport/.

Kentucky Water Resources Research Institute. "Water Distribution System Toolkit". http://www.uky.edu/WDST/SCADA.html.

Kirmeyer, G.J., and AWWA Research Foundation. _Guidance Manual for Maintaining Distribution System Water Quality_. AWWA Research Foundation and American Water Works Association, 2000.

Mamun, A. K., R. F. Islam, R. Haque, G. M. M. Khan, N. A. Prasad, H. Haqva, R. R. Mudliar, and S. F. Mani. "Smart Water Quality Monitoring System Design and Kpis Analysis: Case Sites of Fiji Surface Water". _Sustainability_ 11, no. 24 (2019).

Monroe, West. "3 Enablers to Water Utility Customer Centricity". https://blog.westmonroepartners.com/does-your-water-utility-have-a-customer-centric-workforce/.

Moreno-Munoz, A., F. J. Bellido-Outeirino, P. Siano, and M. A. Gomez-Nieto. "Mobile Social Media for Smart Grids Customer Engagement: Emerging Trends and Challenges". _Renewable and Sustainable Energy Reviews_ 53 (2016/01/01 2016): 1611–16.

Ofwat. "Tapped in – from Passive Customer to Active Participant Report". (2017). https://www.ofwat.gov.uk/publication/tapped-in-from-passive-customer-to-active-participant/.

Oracle. "New Additions! The Best Water Utility Social Media Accounts You Should Be Following Right Now". https://blogs.oracle.com/utilities/the-best-water-utility-social-media-accounts-you-should-be-following-right-now.

Process Technology. "Scada Upgrade Benefits Yarra Valley Water". https://www.processonline.com.au/content/process-control-systems/article/scada-upgrade-benefits-yarra-valley-water-340743931.

PUB. "About the Smart Water Meter Programme". https://www.pub.gov.sg/smartwatermeterprogramme/about.

Public Utilities Board, Singapore. "Managing the Water Distribution Network with a Smart Water Grid". *Smart Water* 1, no. 1 (2016/07/21 2016): 4.

PwC. "Fourth Industrial Revolution for the Earth Harnessing Artificial Intelligence for the Earth". (2018). https://www.pwc.com/gx/en/sustainability/assets/ai-for-the-earth-jan-2018.pdf.

SA Water. "Smart Tech Success in Sa's Sewers". https://www.sawater.com.au/news/smart-tech-success-in-sas-sewers.

_____. "Smart Water Network". https://www.sawater.com.au/current-projects/smart-water-network

Samir, Nourhan, Rawya Kansoh, Walid Elbarki, and Amr Fleifle. "Pressure Control for Minimizing Leakage in Water Distribution Systems". *Alexandria Engineering Journal* 56, no. 4 (2017/12/01/ 2017): 601–12.

Schultz, Wesley, Shahram Javey, and Alla Sorokina. "Smart Water Meters and Data Analytics Decrease Wasted Water Due to Leaks". *Journal – AWWA* 110, no. 11 (2018/11/01 2018): E24–E30.

Silo.AI. "How Artificial Intelligence Is Transforming the Water Sector: Case Ramboll". https://silo.ai/how-artificial-intelligence-is-transforming-the-water-sector-case-ramboll/.

Software Advice. "How to Optimize Utility Asset Management with Gis". https://www.softwareadvice.com/resources/optimize-utility-asset-management-with-gis/.

Southern Water. "Report a Leak". https://www.southernwater.co.uk/help-advice/leaks/report-a-leak

Tsakalides, P., A. Panousopoulou, G. Tsagkatakis, and L. Montestruque. *Smart Water Grids: A Cyber-Physical Systems Approach*. Boca Raton, Florida: CRC Press, 2018.

Tyrväinen, Olli, Heikki Karjaluoto, and Hannu Saarijärvi. "Personalization and Hedonic Motivation in Creating Customer Experiences and Loyalty in Omnichannel Retail". *Journal of Retailing and Consumer Services* 57 (2020/11/01/ 2020): 102233.

Umwelt und Informationstechnologie Zentrum. "Gis in Water Resource Monitoring". http://uizentrum.de/en/gis-in-water-resource-monitoring-2/

United Utilities. "Final Water Resources Management Plan" (2015). https://www.unitedutilities.com/globalassets/z_corporate-site/about-us-pdfs/water-resources/wrmpmainreport_acc17.pdf.

Ventura Water. (2016). https://www.cityofventura.ca.gov/DocumentCenter/View/8309/Pipeline_nov16_print?bidId≥

WaterBriefing. "United Utilities Awards Amp7 Contracts for Supply of Pressure Management Valve Controllers". https://www.waterbriefing.org/home/contracts/item/17094-united-utilities-awards-amp7-contracts-for-supply-of-pressure-management-valve-controllers.

WHO. "Leakage Management and Control – a Best Practice Training Manual". https://www.who.int/docstore/water_sanitation_health/leakage/begin.html#Contents.

WNS. "The Social Media Manual for the Utility Industry: Guidelines, Best-Practices and-Outsourcing-Strategies". https://www.wns.com/insights/articles/articledetail/71/the-social-media-manual-for-the-utility-industry-guidelines-best-practices-and-outsourcing-strategies.

Yorkshire Water. "Open Data". https://www.yorkshirewater.com/open-data/

YSI. "Ysi Technology Used to Monitor Source Water before Reaching Treatment Plants in Nyc". (2016). https://www.ysi.com/File%20Library/Documents/Application%20Notes/A507-Monitoring-Source-Water-Before-It-Reaches-Treatment-Plants-in-New-York-City.pdf.

Anglian Water. "Report a Leak." https://www.anglianwater.co.uk/services/water-supply/leakage/report-a-leak/#:~:text=Such%20is%20our%20success%20in%20tackling%20leaks%2C%20we%27re,years%20ago%20and%20that%27s%20with%2020%25%20more%20customers.

AWWA. "Gamification Boosts Utility's Customer Engagement, Water Efficiency Success." https://www.awwa.org/AWWA-Articles/gamification-boosts-utilitys-customer-engagement-water-efficiency-success.

Esri UK. "Thames Water." (2019). https://resource.esriuk.com/wp-content/uploads/Esri-UK-Thames-Water-14.7.20.pdf.

Government of Dubai. "Dewa's High-Water Usage Alert Service Helps to Reduce 217,370 Tons of Carbon Emissions." https://www.mediaoffice.ae/en/news/2022/August/16-08/DEWAs-High-Water-Usage-Alert-service-helps.

i20. "United Utilities." https://en.i2owater.com/clients/united-utilities/.

ITNews. "Sa Water Boosts Resiliency of Clean Water, Sewerage Services." https://www.itnews.com.au/news/sa-water-boosts-resiliency-of-clean-water-sewerage-services-538538.

____. "Sydney Water Trials Acoustic Sensors for Leaks, Saves 9000 Megalitres." https://www.itnews.com.au/news/sydney-water-trials-acoustic-sensors-for-leaks-saves-9000-megalitres-580664.

PUB. "Subscribe to Sms Alerts." https://www.pub.gov.sg/drainage/floodmanagement/subscribesms.

Robert C. Brears. "Autonomous Drones Monitoring Singapore's Reservoirs." Mark and Focus, https://medium.com/mark-and-focus/autonomous-drones-monitoring-singapores-reservoirs-4abd3cddb622.

____. "Tackling the Water Crisis: San Diego's 'Waste No Water' Smartphone App Leads the Way in Conservation." Mark and Focus, https://medium.com/mark-and-focus/the-smart-app-to-waste-no-water-4375684b13bc.

SA Water. "Continued Sensor Success for Sa Water's Smart Network." https://www.sawater.com.au/news/continued-sensor-success-for-sa-waters-smart-network.

United Utilities. "App." https://www.unitedutilities.com/app/.

Watercare. "Tap into Our App." https://www.watercare.co.nz/Help-and-advice/Smart-meter-app.

Xylem. "Nyc Monitors Source Waters with Ysi Buoys." (2021). https://www.ysi.com/file%20library/documents/application%20notes/a507-monitoring-source-water-before-it-reaches-treatment-plants-in-new-york-city.pdf.

Yorkshire Water. "Promising Results for Ai Leak Detection Trial." https://www.yorkshirewater.com/news-media/news-articles/2021/yorkshire-water-sees-promising-results-in-ai-leak-detection-trial/.

# Chapter 8
# Innovative financial instruments and approaches for water projects

**Abstract:** Global demand for water is increasing due to population growth, urbanisation, economic development, and changing consumption patterns. Climate change will increase the numbers of people exposed to both flooding and droughts. Meanwhile, various climatic and non-climatic trends will reduce water quality in waterways, affecting both humans and nature. However, there is significant underinvestment in water resources management, including nature-based solutions, to meet these challenges globally. Nonetheless, there are various innovative financial instruments and approaches available to ensure the provision of sustainable, reliable, resilient, and affordable water and water-related services that meet customers' expectations in the future.

**Keywords:** Water Financing, Economic and Financial Instruments, Tradable Permits, Payment for Ecosystem Services, Green Bonds, Public-Private Partnerships

## Introduction

Global demand for water is increasing due to population growth, urbanisation, economic development, and changing consumption patterns. Climate change will increase the numbers of people exposed to both flooding and droughts. Meanwhile, various climatic and non-climatic trends will reduce water quality in waterways, affecting both humans and nature. However, there is significant underinvestment in water resources management, including nature-based solutions, to meet these challenges globally.[1,2,3] This chapter first discusses the initial lack of financing available for water projects before discussing the various innovative financial instruments and approaches available to ensure the provision of sustainable, reliable, resilient, and affordable water and water-related services that meet customers' expectations in the future.

## 8.1 Overcoming barriers to water financing

Globally, there needs to be an additional investment of $1.7 trillion to ensure universal and equitable access to safe and affordable drinking water for all, which is around three times the current investment levels. Meanwhile, the scale of investments in water security needs to increase significantly, with estimates ranging from $6.7 trillion by 2030 to $22.6 trillion by 2050. Nonetheless, there are a few barriers that create a gap between current financing and future needs, including the following:

https://doi.org/10.1515/9783111028101-008

- Water is generally an under-valued resource, not properly accounted for by investors that depend on or affect its availability in other sectors such as urban development, agriculture, and energy etc.
- Water services are often under-priced, resulting in low cost-recovery for water investments
- Water infrastructure is generally capital intensive, with high sunk costs and long pay-back periods
- Water management provides both public and private benefits, many of which cannot be easily monetised. This reduces potential revenue flows
- Water projects are often too small or too context-specific, raising transaction costs and making innovative financing models difficult to scale-up
- Business models often fail to support operation and maintenance efficiency, hampering the ability to sustain service at least cost over time[4,5,6,7]

To overcome these barriers, the High Level Panel on Water has defined a range of principles that should be followed to help finance investments, enhance water services, mitigate water-related risks, and contribute to sustainable growth:
- *Maximise the value of existing assets for water-related investments*: Service providers can reduce overall investment needs and improve capital efficiency through improving the operational efficiency and effectiveness of existing infrastructure. Improvements can result from good operation and maintenance of infrastructure and demand management
- *Design investment pathways that maximise water-related benefits over the long-term*: The multiple benefits that water-related investments generate depend on how investments are designed and sequenced to meet strategic goals, including climate change adaptation. This means projects should be designed to be scalable and adjustable to changing conditions
- *Ensure synergies and complementarities with investments in other sectors*: Policies outside of the water sector should be encouraged to factor in water risks, which in turn stimulates water-wise investments
- *Attract more financing by improving the risk-return profile of water investments*: Governments can employ a range of fiscal policy instruments to recover the costs of investments from beneficiaries, improve the financial viability of utilities, and provide a revenue stream to improve the risk-return profile of water-related investments[8]

### 8.1.1 Economic and financial instruments

Economic instruments enable environmental or social costs to be incorporated into the price of goods, services or activities that give rise to them. This sends a price signal to users or consumers to reduce inefficient and wasteful use of resources and foster their optimal allocation. Economic instruments can be used to promote eco-efficient economic activities, therefore promoting innovation and competitiveness. Regarding pollution, economic instruments enable the implementation of the polluter-pays principle, making the polluter instead of society as a whole pay for the damage they cause. At the same time, ensuring the provision of sustainable, reliable, resilient, and affordable water and water-related services will require significant investments and therefore financing.[9,10]

## 8.2 Water prices

Water prices have the primary goal of financing water supply infrastructure. The price of water should be set at a level that ensures the recovery of costs for each sector (agriculture, households, and industry) and the allocation of costs to each sector (avoiding cross-subsidies). Water prices should relate to three types of cost:

- *Direct economic costs*: Full recovery of the economic costs of the water services requires the water price to include:
  - Operational and maintenance costs of water infrastructure
  - Capital costs for the construction of this water infrastructure
  - Reserves for future investments in water infrastructure
- *Social costs*: The social costs, direct and indirect, of providing water services varies mainly with respect to specific contextual settings. As such, calculating and comparing these costs across different settings is generally not feasible
- *Environmental costs*: The environmental costs of economic activities are generally not reflected in the prices established in the market but appear as externalities. The principle of full cost recovery requires that these costs be taken into consideration

Overall, water prices which represent full costs (economic and environmental costs) provide price signals to water users to be more efficient while generating the means for ensuring sustainable water infrastructure.[11,12]

---

**Case 8.1: San Diego' New Water Billing Rates**

San Diego's Public Utilities Department has issued new water billing rates effective January 1, 2023. The rates vary based on the type of customer, with single-family domestic customers having a typical 3/4-inch meter, although some larger homes may have a 1-inch meter. The total bill for a single-family domestic customer is a combination of the monthly meter base fee and the amount of water used. Water usage is measured by hundred cubic feet (HCF), with each HCF equalling 748.05 gallons. The monthly charges for single-family domestic customers are based on a tiered system, with the base fee set at $27.77. The rates increase as the amount of water used increases, with the first 4 HCF used billed at $5.550 per HCF, the next 8 HCF used billed at $6.217 per HCF, and the following 6 HCF used billed at \$8.881 per HCF. Each HCF used after the initial 18 HCF is billed at \$12.488 per HCF. Other customer types, including multi-family domestic customers, commercial and industrial customers, temporary construction customers, and irrigation customers, have their own monthly meter base fees and rates per HCF of water used.[13]

## 8.3 Stormwater fee discounts

Cities can incentivise developers and property owners to manage stormwater as well as preserve open space and protect or plant trees by offering stormwater fee discounts. Stormwater fee discounts aim to reduce the required capacity and cost of stormwater treatment practices. They allow property owners to reduce the amount of stormwater fees they pay by decreasing impervious surfaces or by using green infrastructure techniques that reduce the amount of stormwater runoff. Before setting the discount, the agency in charge should set appropriate management goals and determine how to credit private property owners for the action being incentivised. It is common for cities to provide a percent discount for the level of performance. This discount is usually given for stormwater quantity reductions. Discounts can also be offered for impervious surface reductions or for implementing specific green infrastructure practices.[14,15,16]

**Case 8.2: Stormwater Service Credits for Businesses in the City of Guelph**
Industrial, commercial, institutional, and multi-residential properties in the City of Guelph, Canada, may be eligible for stormwater fee credits of up to 50% to help reduce runoff and protect the city's water supply and wildlife habitat. To qualify for the credit, the property must have six units or more. The Stormwater Credit Program has four categories of credits, which align with the objectives of the city's stormwater management programme. The categories include peak flow reduction, runoff volume reduction, water quality treatment, and operations and activities. The maximum credit available for any property is 50% and is capped accordingly (Table 8.1). The credit amount is based on the eligible stormwater measures and property eligibility as outlined in the Stormwater Credit Application Guidance Manual. The manual provides an overview of the Stormwater Credit Program framework, the application process, site inspection requirements, and penalties for violations of the terms and conditions. To apply for a stormwater fee credit, property owners need to complete the Stormwater Credit Application Form and submit it to the city for review. The credit application form provides instructions and information on the required documentation and verification of the eligible stormwater measures. Property owners can use this credit to reduce their stormwater service fees while contributing to the protection of Guelph's water resources and the environment.[17]

**Table 8.1:** City of Guelph's Stormwater Credit Categories.

| Credit Category | Description | Maximum Credit (capped at 50 percent) |
| --- | --- | --- |
| Peak flow reduction | Facilities that control the peak flow of stormwater discharged from the property based on the outlet rate in comparison to natural hydrologic conditions | 15 percent |
| Runoff volume reduction | Facilities that control the amount of stormwater retained on the property, based on retention volume resulting from increased infiltration, evapotranspiration, or reuse | 40 percent |
| Water quality treatment | Facilities that control the quality of stormwater discharged from the property, based on treatment type, pollutant load reduction, or Ministry of the Environment and Climate Change Resources level of protection | 15 percent |
| Operations and activities | Non-structural measures including education programmes and pollution prevention/risk management practices | 15 percent |

## 8.4 Stormwater volume credit trading

Stormwater volume credit trading provides an onsite compliance option for property developers or owners who are subject to stormwater management regulations. In many cases, regulations include onsite retention or detention requirements for new developments or redevelopment projects over a certain size. A credit trading programme enables developers or property owners subject to these regulations to meet all or a portion of their requirements offsite by buying volume-based stormwater credits. These credits are generated from the installation and maintenance of green infrastructure projects located offsite. Specifically, credits can be generated by:

- Developers or property owners who voluntarily implement green infrastructure retrofit projects on properties that are not subject to post-construction green infrastructure requirements
- Developers or property owners who are subject to green infrastructure requirements but build green infrastructure projects that exceed minimum stormwater requirements

A trading programme requires a local entity to oversee and manage the trading marketplace and ensure that the green infrastructure projects that generate the credits are properly maintained over time. This function is usually provided by a stormwater agency, but an independent entity can be created to administer the programme.

Some of the main benefits of a stormwater volume credit programme include the following:

–  It allows flexibility for developers and property owners as they can choose the cheaper option. In some cases, it will be cheaper to buy credits from an offsite provider than managing stormwater onsite
–  Buying credits can allow developers or property owners to make use of additional buildable areas onsite, including rooftop or underground areas
–  Overall water quality in a city or watershed can improve as a trading programme allows for a greater number of small green infrastructure installations across an area in comparison to a smaller number of larger green infrastructure practices, all onsite[18]

**Case 8.3: Washington, DC's Stormwater Retention Credit Trading Program**
The Stormwater Retention Credit (SRC) Trading Program allows property owners in the District of Co-lumbia to generate and sell SRCs to earn revenue for projects that reduce harmful stormwater runoff by installing green infrastructure or removing impervious surfaces. The program is designed to en-courage voluntary green infrastructure projects in the Municipal Separate Storm Sewer System (MS4) that generate High-Impact SRCs, which provide the greatest water quality benefits to the District's waterbodies. Property owners can sell SRCs to the Department of Energy and Environment (DOEE) through the SRC Price Lock Program or in an open market to properties that have regulatory require-ments for managing stormwater. DOEE provides resources to help property owners get started on an SRC-generating project, which can be done on their property or by aggregating SRC from multiple properties. The SRC Price Lock Program enables eligible SRC generators to sell SRCs to DOEE at fixed prices, with the option to sell to another buyer. The program aims to make it easier to generate SRCs on land owned by non-profits, and DOEE prioritizes funding for these projects. The application process for the SRC Trading Program is available year-round, but DOEE plans to host biannual, 60-day open application windows for funding opportunities in the spring and fall. Property owners can submit their applications online through the DOEE Surface and Groundwater System, and eligibility requirements and steps for participation are available in the program guide. The SRC Trading Program provides an opportunity for property owners to earn revenue while contributing to the protection of the District's water resources and environment.[19]

## 8.5 Environmental taxes

According to the OECD, environmental taxes directly address the market failure that causes markets to ignore environmental impacts. A well-designed environmental tax increases the price of a good or activity to reflect the cost of the environmental harm that it imposes on others. The harm to others, an externality, is internalised into mar-ket prices. This ensures that economic actors, including consumers and firms, take these costs into account in their decision-making. Environmental taxes provide other benefits including an ongoing incentive to abate (they provide a continuous incentive for the abatement of pollution), improving the competitiveness of environmentally-friendly alternative technologies (helping make alternatives more viable without the need for direct subsidies), and providing a strong incentive to innovate (taxes in-crease the cost to a polluter of generating pollution, providing firms with an incentive to develop innovations and adopt existing ones). Meanwhile, an alternative to environmental taxes is the providing of tax incentives to subsidise environmentally beneficial goods or actions. In the context of water resources management, environmental taxes are compulsory payments to fiscal authorities for behaviours that lead to the degradation of the water environment, with the objective being to encourage alternative behaviour to the one targeted by the tax, for example, the use of less polluting techniques and products. Overall, environmental taxes should:

- Be designed to target the pollutant or polluting behaviour with few exceptions
- Be as broad as the scope of the environmental damage
- Be commensurate with the environmental damage
- Be credible and its rate predictable to motivate environmental improvements
- Assist fiscal consolidation or help reduce other taxes
- Be clearly communicated to ensure public acceptance[20,21]

**Case 8.4: Denmark's pesticide tax**

Denmark is one of the world's most heavily farmed countries, and its agricultural sector has been undergoing significant structural changes in recent years towards fewer but larger farms. However, this has also led to high environmental pressure from the agricultural sector. To address this, Denmark has carried out an active pesticide policy since 1986, with the aim of reducing agricultural pesticide use by 50%. One of the key tools in this policy was the introduction of a value-added tax on pesticides, but it was far less effective than predicted. Thus, a new tax was introduced in 2012, designed to target the health and environmental impacts of pesticides more directly. This new tax differentiated tax rates according to the health and environmental load of each pesticide, offering a farmer an incentive to substitute more harmful pesticides with less harmful ones. The pesticide load indicator is based directly on a risk assessment and combines three sub-indicators for human health, ecotoxicology, and environmental fate. The tax rate is 107 DKK (14.30 EUR) per pesticide load plus a basic tax of 50 DKK (6.70 EUR) per kg of active ingredient. This new tax design resulted in substantial variation in tax levels across pesticides due to large variation in environmental load. An analysis of the list price of 38 pesticides before and after the tax change showed that changes in taxes have largely been reflected in the market price of pesticides. The new tax policy was accompanied by a requirement for farmers to submit their electronic spray records to the Ministry of Environment and Food, offering more valid and reliable data on pesticide use than sales figures. The revenue from the tax is reimbursed to the agricultural sector through a reduction in land taxes. The policy objective was set as a 40% reduction in the Pesticide Load Indicator between 2011 and 2015, based on sales rather than reported use.[22]

## 8.6 Subsidies

The OECD defines subsidies as *"government interventions through direct and indirect pay amends, price regulations, and protective measures to support actions that favour environmentally unfriendly choices over environmentally friendly ones"*. Nonetheless, subsidies are economic instruments that can be used as an incentive to stimulate change in user behaviour towards environmentally friendly conduct or encourage investments in environmentally friendly production techniques, mitigating or eliminating adverse effects.[23] The following principles should be followed to ensure subsidies promote environmentally friendly conduct or technologies:

– *Subsidies should achieve the intended policy outcome*: Subsidies require a smart design and clarity about what the policy objectives and short- and long-term objectives are
– *Subsidies should reach the intended target groups*: They require clarity on who is the intended target group and how they can best be reached. It also requires rigorous monitoring to track how subsidies are reaching the intended groups
– *Subsidies should be financially sustainable*: A thorough understanding of the potential costs of the programme is required. Costs include both upfront capital costs and long-term operational and maintenance costs
– *Subsidies should integrate local peoples' needs*: To guarantee the sustainability of the subsidised environmental technology, it is of prime importance to facilitate the integration and participation of the local beneficiaries and to develop a sense of ownership towards the new infrastructure
– *Subsidies should be implemented clearly and transparently*: As subsidies involve public funds, subsidy programmes need to be clear and transparent, enabling eligible households or communities to access them and providing clear recourse mechanisms in cases where there is a suggestion of impropriety[24]

---

**Case 8.5: Flanders' subsidies for environmentally friendly investments**

Flanders, Belgium, offers various ecological investment support options for companies interested in going green. Flanders Innovation & Entrepreneurship (VLAIO) provides grants and other financial aids to companies investing in environmental technologies, energy technologies, and renewable energy and combined heat and power. The Ecology Premium Plus (EP-Plus) is a grant available to companies investing in technologies listed on a limitative technology list (LTL). The LTL includes about 30 technologies in three categories: environmental technologies, energy-saving investments, and renewable energy and cogeneration. The EP-Plus grant is not applicable for ecological investments eligible for aid through green certificates and cogeneration certificates. The amount of the EP-Plus grant varies according to the size of the enterprise and the technological category. The Strategic Ecology Support (STRES) is an additional funding tool for companies using technologies that are not listed on the LTL but are being deployed in strategic environmental projects. The ecology bonus amounts to 20%, 30%, or 40% of the additional costs, depending on the enterprise's scope and technology. The STRES support for each company is limited to a maximum of €1 million every three years. The ecological investment support in Flanders aims to encourage companies to organize production processes in a more eco-friendly and energy-efficient way by funding part of the additional costs involved in such investments. The eco-class and eco score of a technology determine the subsidy percentage and the maximum amount of support a company can receive.[25]

---

## 8.7 Tradable permits

With a variety of climatic and non-climatic trends resulting in water scarcity and pollution, a range of locations have implemented market mechanisms based on consumption rights and pollution in the management of water resources. Specifically, tradable permits are one of the most efficient market-based instruments for allocating water resources and for mitigating pollution of water resources.[26] There are two main tradable permit systems: Tradable water abstraction rights and tradable water pollution rights.

## 8.7.1 Tradable water abstraction rights

These rights are for quantitative water resource management with water rights being either permanent and unlimited (property rights to the water resource) or temporary and limited (transferable rights to use water without right of abuse). In a tradable water abstraction rights regime, the water authority sets a water consumption cap, which is the maximum amount of water that can be abstracted. It allocates the abstraction rights among the basin users, who then can exchange them based on their present and/or future expected water consumption demand. Water users are encouraged to use water efficiently for two reasons. First, it reduces the need to purchase costly abstraction rights and second, they can gain revenue from selling excess water rights once they reduce their water consumption.

## 8.7.2 Tradable water pollution rights

Tradable water pollution rights are used for the protection and management of water quality. The water management authority establishes the maximum amount of emissions according to the carrying capacity of the ecosystem that is focussed on. The total amount of emissions is divided into a fixed number of permits or rights to pollute, which can be initially allocated to economic actors according to their past levels of pollution, known as grandfathering, or via auction. The holders can then trade the rights purchased in a secondary permit market. This means a polluting point source, which has low abatement costs, can sell permits to sources with high clean-up costs. The result is the total cost of reducing pollution is minimised as pollution-reduction efforts are carried out by economic actors who can do it at the lowest cost.[27,28]

**Case 8.6: Tradable water abstraction rights: Fox Canyon Water Market**
Ventura County, California, generates $2.1 billion from agriculture. At the same time, there is significant population pressure with around 450 people per square mile: around five times the average population density of the United States. With groundwater being a critical resource, the state passed the Sustainable Groundwater Management Act (SGMA) in 2014 to ensure the future sustainability of groundwater supplies. Following the passage of SGMA, The Nature Conservancy (TNC) applied for a Conservation Innovation Grant from the U.S. Department of Agriculture (USDA) to develop the Fox Canyon Water Market. The grant enabled TNC to provide support to the Fox Canyon Groundwater Management Agency and project partner California Lutheran University in their effort to establish a

market-driven, producer-led approach to reducing groundwater pumping. Under a cap-and-trade-like system, agricultural producers in the Fox Canyon area are subject to fixed groundwater allocations based on historical use. Producers can then purchase or sell their unused allocation. The market is online, anonymous, and uses an algorithm-driven matching platform, resulting in a level playing field and a fairer deal for farming operations of all sizes. After years of development, the market exchange opened in March 2020. Still in its pilot phase, the market has already seen 58-acre feet of pumping allocations change hands. The successful launch of the pilot is due to a variety of factors including:
- *Water scarcity*: Water scarcity requires innovative solutions
- *Fixed allocations*: Participating producers voluntarily agreed to a fixed allocation of groundwater
- *Agricultural stakeholder support*: From the beginning, the project has been collaborative, and producer-driven
- *Market design expertise*: The project leveraged the experience of TNC and partners in designing environmental markets, including robust pilot testing
- *Capacity and funding*: TNC provided robust planning and oversaw stakeholder engagement efforts to provide expertise and gain support from the USDA[29]

**Case 8.7: Tradable water pollution rights: Ohio River Basin Trading Project**
The Ohio River Basin Trading Project is the United States' first interstate trading plan signed by Ohio, Indiana, and Kentucky in 2012, making it the world's largest water quality trading programme. The project, which has been extended through 2020, achieves water quality goals by allowing permitted dischargers to purchase nutrient reductions from another source. The cost of reducing nutrient discharges can differ from one emitter to another, and water quality trading provides an option for meeting discharge requirements in a cost-effective manner. The project is voluntary with the incentive to participate based on credit sellers receiving attractive financial benefits from the selling of credits and the permitted dischargers having the flexibility to cost-effectively meet their environmental permit requirements. Measures of success during the pilot include:
- Identifying and overcoming barriers to successful full-scale roll-out
- Implementing trading mechanisms that are ecologically effective and acceptable to participants and other stakeholders
- Promoting early, voluntary participation
- Measuring the extent to which the broader ecosystem services can be supported through the project
- Establishing the full suite of systems and protocols needed for a complete and compliant programme[30]

## 8.8 Payment for watershed ecosystem services

A Payment for Ecosystem Services (PES) scheme is defined as a voluntary, conditional agreement between at least one seller and one buyer over a well-defined ecosystem service, or land use that is assumed to produce that service. Globally, there are hundreds of PES initiatives that provide direct payments to landowners for undertaking specific land-use practices that can increase the provision of biodiversity conservation, prevent erosion, enhance carbon sequestration, and improve scenic beauty, as well as provide other ecosystem services that are of interest, directly or indirectly, to humans. PES initiatives must achieve the same level of environmental benefits at a lower cost than other possible policies to be effective.[31,32] Payment for Watershed Ecosystem Services (PWES) have been widely implemented in both developed and developing countries at different scales to resolve upstream-downstream conflicts, with the most common focus being on water quality, water quantity, and flow regulation.[33] For PES and PWES schemes to be both environmentally beneficial and cost-effective, the OECD recommends a set of 12 criteria to be followed when these schemes are developed:

1. *Remove perverse incentives*: For a scheme to be clear and effective, there should be no conflicting market distortions, such as environmentally harmful subsidies available
2. *Clearly define property rights*: The individual or community whose land-use decisions affect the provision of ecosystem services must have clearly defined and enforceable property rights over the land
3. *Clearly define goals and objectives*: The scheme should have clear goals to help guide the design of the programme, enhance transparency, and avoid political influence
4. *Develop a robust monitoring and reporting framework*: The monitoring and reporting of biodiversity and ecosystem services is fundamental, enabling the assessment of the programme's performance and improvements over time
5. *Identify buyers and ensure sufficient long-term financing*: There must be sufficient and sustainable financing of the scheme to ensure its objectives can be achieved
6. *Identify sellers and target ecosystem service benefits*: Payments should be prioritised to areas that provide the highest benefits
7. *Establish baselines and target payments to ecosystem services that are at risk of loss, or to enhance their provision*: The programmes should only make payments for ecosystem services that are additional to the business-as-usual baseline, in other words in the absence of the programme
8. *Differentiate payments based on the opportunity costs of ecosystem service provision*: Programmes that reflect ecosystem providers' opportunity costs via differentiated payments can achieve greater aggregate ecosystem service provision per unit cost
9. *Consider bundling or layering multiple ecosystem services*: Joint provision of multiple services can increase the benefits of the programme while reducing transaction costs, especially if finance for multiple benefits is available
10. *Address leakage*: Leakage occurs when the provision of ecosystem services in one location increases pressure for conversion in another. If leakage is potentially high, the monitoring and accounting framework can be expanded to enable assessment of potential leakage so measures can be implemented to address it
11. *Ensure permanence*: Events, such as disasters, may undermine the ability of a landowner to provide an ecosystem service as stipulated in the agreement. If the risks are high, this will impede the effective functioning of the market, and so insurance mechanisms need to be considered

12. *Deliver performance-based payments and ensure adequate enforcement*: Payments should be ex-post, conditional on ecosystem performance. If this is not possible, effort-based payments, for example, improvements in management practices, are a suitable alternative. There should also be disincentives for breaching the agreement which is enforced[34]

---

**Case 8.8: Milwaukee River Pay-for-Performance Project**

Over the period 2013–2017, the Milwaukee River Pay-for-Performance Project, funded by the Great Lakes Protection Fund, rewarded farmers for improving water quality by reducing phosphorous loss from agricultural land in the West Branch of the Milwaukee River, a 58 square-mile Wisconsin watershed. Farmers were rewarded according to science-based outcomes where the environmental impacts of farmer-selected conservation practices were tracked, and payments made based on verified environmental improvements. A pay-for-performance approach was taken as it is data-driven and science-based, results in measurable water quality improvements, is cost-effective for farmers and conservation programmes, provides flexibility and allows for farmer innovation, and expands market opportunities for water quality trading. The project had five steps:

1. Farmers implement the most appropriate and cost-effective strategies for their farm
2. Science-based models predict less phosphorous entering the stream
3. Farmers are paid based on modelled farm-level results
4. Water quality improvements are monitored and verified
5. Farmers are paid based on monitored watershed-level results

The farmers were paid by wastewater treatment plants or other downstream entities that needed to meet water quality obligations as well as conservation programmes that wished to show measurable water quality outcomes. Overall, by changing field management practices, participating farmers reduced phosphorous losses by as much as 40 percent.[35,36]

---

## 8.9 Green bonds

Green bonds can help mobilise resources from domestic and international capital markets for climate adaptation and other climate and environmentally friendly projects. They are like conventional bonds, but the proceeds are invested in projects that generate climate and environmental benefits such as sustainable land use, biodiversity, and clean water.[37] To ensure green bonds genuinely contribute to climate and environmental targets, the following initiatives have been developed.

### 8.9.1 The Green Bond Principles

The Green Bond Principles (GBP), established by the International Capital Market Association, are voluntary process guidelines that recommend transparency and disclosure and promote integrity in the development of the green bond market by clarifying the approach for issuance of a green bond. The GBP recognises several categories of eligible green projects that contribute to the climate and environment, including climate change mitigation, climate change adaptation, natural resource conservation, biodiversity conservation, and pollution prevention and control. With regards to water management, the GBP provides an overview of eligible projects including sustainable infrastructure for clean drinking water, wastewater treatment, sustainable urban drainage systems, and river training and other forms of flood mitigation.[38]

## 8.9.2 Labelling scheme for green bonds

The Climate Bonds Initiative's Climate Bonds Standard and Certification Scheme is a labelling scheme for bonds. The scheme uses a rigorous scientific criterion to ensure the green bonds are consistent with the Paris Agreement and is used by bond issuers, governments, investors, and financial markets to prioritise investments that make genuine contributions to addressing climate change. The new Water Infrastructure Criteria has been developed that lays out the requirements that water infrastructure assets and/or projects must meet to be eligible for inclusion as a Certified Climate Bond, as summarised in Table 8.2.[39]

**Table 8.2:** The Water Infrastructure Criteria.

| Step | | Description | | |
|------|---|-------------|---|---|
| 1. | Comply with mitigation component | Greenhouse gas emissions from water projects are not to increase. Instead, they are to comply with business-as-usual baselines or aim for emission reduction over the operational lifetime of the water asset or project | | |
| 2. | Comply with adaptation and resilience component | The water infrastructure and its surrounding ecosystem are resilient to climate change and have sufficient adaptation to address climate change risks. To demonstrate this, the issuers should complete a scorecard made up of five sections: | 1. | *Allocation*: Addressing how water is shared by users within a given basin or aquifer |
| | | | 2. | *Governance*: Addressing how/whether water will be formally shared, negotiated, and governed |
| | | | 3. | *Technical diagnostic*: How/whether changes to the hydrologic system are addressed over time |
| | | | 4. | *Nature-based solutions*: For nature-based and hybrid infrastructure, the issuers need to have sufficient understanding of ecological impacts at/beyond project site with ongoing monitoring and management capacity |

**Table 8.2** (continued)

| Step | Description |
|------|-------------|
| | 5. *Assessment of the adaptation plan*: There needs to be a check of how complete the coping mechanisms are to identify climate vulnerabilities |

**Case 8.9: DC Water's green bonds**

In July 2014, DC Water issued its inaugural green bond to finance a portion of the DC Clean Rivers Project. This historic $350 million issuance represented DC Water's inaugural green bond issue and the first "certified" green bond in the US debt capital markets with an independent second party sustainability opinion. It was also the first municipal century bond issued by a water/wastewater utility in the United States. The issuance achieved its green certification based upon the DC Clean Rivers Project's environmental benefits, which include improving water quality by remediating combined sewer overflows, promoting climate resilience through flood mitigation and improving quality of life through promotion of biodiversity and waterfront restoration. DC Water continues to offer green bonds, attracting diverse investors including a new class of socially and environmentally conscious investors. The DC Water Board has adopted a Green Bond Framework which is aligned with the four ICMA (International Capital Market Association) Principles regarding use of proceeds, project selection, management of proceeds, and reporting. The framework formalizes the process and commitments governing the issuance of Green Bonds by the District of Columbia Water and Sewer Authority (DC Water). It aligns with the four pillars of the Green Bond Principles: Use of Proceeds, Project Evaluation and Selection Process, Management of Proceeds, and Reporting. Proceeds from Green Bond issuances may be used to fund the Clean Rivers Project, which encompasses pollution prevention and control, sustainable water and wastewater management, and climate change adaptation. Additionally, proceeds may be used to finance other projects that fall under various Green Project Categories, subject to Board approval. Management of proceeds involves depositing the net proceeds of Green Bond issuances in a segregated account, holding funds exclusively in US Treasury securities or bank deposits, and allocating all proceeds to eligible project expenses within three years. DC Water will publish an annual Green Bond Report, providing transparent communication on the use of proceeds, environmental and social outcomes achieved, and responsible management of the project and DC Water. The report will be available on DC Water's website and the Electronic Municipal Market Access website. The Clean Rivers Project and other projects financed by Green Bonds will have appropriate performance measures for environmental, social, and governance factors associated with the projects. DC Water will seek an independent Second Party Opinion on the sustainability of each Green Bond issuance.[40]

## 8.10 Public-private partnerships

Public-private partnerships (PPP) are long-term, contractual agreements between a public entity and a private operator/company (or a consortium), under which a service is provided. PPPs involve a process where private operators bid for a contract to design, finance, and manage the risks involved in delivering public services or assets.

In return, the private contractor receives fees from the public body and/or user charges for the long-term operation and maintenance of the asset. There are two types of green PPP projects:
- *Greenfield projects*: These projects develop new infrastructure, such as a new wastewater treatment plant
- *Brownfield projects*: In these projects, the private sector participates (as investors and operators) in existing infrastructure facilities[41]

For PPPs to be successful, a range of conditions need to be met:
- *Effective partnerships*: Effective partnerships are crucial to the success of PPPs. Unlike traditional procurement for assets or services, which use short-term contracts to acquire or renovate public assets, a PPP is a global contract that may last anywhere from 15 to sometimes more than 90 years. As such, establishing a real partnership based on cooperation, expertise, and credible commitment is essential. Also, the public body must acquire internal knowledge and expertise necessary to define the terms of the agreement
- *Interaction and negotiation*: Interaction and negotiation with an operator or operators during the call for bidders' phase can clarify objectives of the partnership and provide innovative technological solutions not yet envisioned by the public body. This is helpful for PPPs negotiated in an uncertain environment with complex technologies that vary in speed of obsolescence. For this phase to be successful, the public body needs to invest in gaining expertise and generating enough competition to challenge private operators
- *Clear environmental objectives*: Clear environmental objectives and their weights in the procedure to award PPP projects need to drive effective environmental-related PPPs. The addition of green requirements to the project specification after PPP design will be costly and likely incompatible with technological choices put in place. Also, environmental targets need to be measurable and clearly defined, with agreed-upon approaches for *ex-post* monitoring
- *Flexibility*: Flexibility is a crucial element of PPPs. Discussion with private operators for a PPP should focus on efficient and flexible solutions that allow for a quick response to changing requirements and new technologies. Also, the contract should describe and anticipate how the relationships evolve over time as soon as unanticipated events occur[42]

**Case 8.10: The largest public-private partnership for wastewater operations in the United States**
The City of Wilmington, Delaware, has selected an international engineering firm to operate and manage its wastewater treatment plant, combined sewer overflow facilities, and its Renewable Energy Biosolids Facility. The city is Delaware's largest with a wastewater operation serving more than 400,000 residents. The agreement, which combines the operations and maintenance of all facilities under the engineering company's management, has provisions for additional engineering studies and design-

build projects to renew existing structures and develop value-added projects. The City estimates the base contract is valued at $20 million per year for an initial 20-year term, with options for two additional two-year extensions, for a possible contract term of 24 years. With the wastewater treatment plant having a maximum design flow of 168 million gallons per day (MGD) and up to 320 MGD in wet weather, the contract is one of the largest PPPs for wastewater operations in the United States. As part of the contract, the engineering company will improve the plant's performance and ensure it becomes a net-zero energy facility that reduces greenhouse gas emissions.[43]

### 8.10.1 Public-private partnerships for ecosystem restoration

Existing financial, legal, and policy mechanisms of PPPs are suitable for major ecosystem restoration initiatives, including protecting and restoring the health of rivers, improving water quality, restoring and enhancing significant areas of habitat, and sequestering significant quantities of carbon dioxide. PPPs are suitable for major ecosystem restoration initiatives as they provide a suitable framework for:
- Sharing costs and benefits between governments, investors, businesses, and the environment
- Leveraging private sector skills, capacity, and capital with strategically directed public funds
- Generating new models of achieving ecosystem restoration outcomes and improving cost effectiveness of ecosystem restoration programmes
- Combining public and private sector knowledge, skills, land, and capital in ways that could result in landscape-wide benefits
- Turning large scale landscape change into business opportunities, creating new asset classes

For the PPPs in ecosystem restoration to be successful, there needs to be:
- Measurable performance standards developed
- Government payments that flow to projects sequentially when they meet specified environmental service standards
- A relatively long-term commitment with the term depending on the nature of the project
- Either a dedicated party responsible for contracting with multiple providers or a government agency contracting directly to multiple providers
- Risk allocation defined at the contracting phase
- Outcomes monitored to determine the delivery of the specified ecosystem services and accompanying payment
- Where possible, project costings will be determined by competitive tendering processes or equivalent[44]

**Case 8.11: The United Kingdom's Natural Environment Impact Fund partnering with the private sector**

The Government of the United Kingdom's 25 Year Environmental Plan has made clear that while the public sector will continue to be an essential source of funding for the natural environment, this must be alongside private sector investment to protect and enhance the environment. In support, the government has committed £10 million in the Budget, from 2021, to support natural environment projects that attract private sector investment through the Natural Environmental Impact Fund. As part of this, the Department for Environment, Food and Rural Affairs (Defra), Esmée Fairbairn Foundation (EFF), and Triodos Bank UK have formed a collaborative partnership to encourage private sector investment in environmental projects that help tackle climate change and restore nature. Four projects that protect and restore valuable habitat have been selected to receive funding in a pilot scheme to encourage sustainable private sector investment in the natural environment. The projects, having been sourced and evaluated by Triodos Bank UK, will receive grant funding from Defra, the EA, and EFF to support their development, complete business plans to attract private sector investment, and deliver long-term environmental benefits and sustainable financial returns. The four projects are:

–   *Devon Wildlife Trust's restoration of the Caen wetlands*: The wetlands site is one of the UK's most important sites for wetland birds but is under pressure from human impacts, climate change, and rising sea levels. The project is a bold and innovative proposal to create a stunning habitat and visitor resource in northern Devon. Alongside the restoration of the habitat, the development of the site for ecotourism through a visitor centre and other facilities will provide a source of income, with the seed funding being used by Devon Wildlife Trust to develop a business case for investment in this project
–   *Rivers Trust's work on natural flood management in the Wyre catchment in Lancashire*: Hard engineering alone will not address future flood risk challenges, and natural solutions must supplement them. The seed funding will allow The Rivers Trust to work with the Wyre Rivers Trust, Environment Agency, United Utilities, Triodos Bank UK, Co-op Insurance, and Flood Re, to develop a financial instrument that would allow upfront investment from the private sector to be reimbursed by the beneficiaries of a healthier environment
–   *National Farmers Union's (NFU) work to reduce nitrate pollution in Poole Harbour*: One of the largest natural harbours in the world, Poole Harbour in Dorset is of international importance to wildlife. However, it is under pressure, with nutrients such as nitrate from agriculture in its catchment flowing down into the harbour and leading to a rapid growth of algae which smothers the estuarine habitat and reduces the amount of food available for birds. Through the Poole Harbour Nutrient Management Scheme, the NFU is aiming to work with and support the farmer-led collaboration in the catchment, equipping them with tools to reduce their use of nitrates. The proposed innovative, industry-led model is designed to offer both environmental benefits and productivity gains for farming businesses while the local community, water companies, and local government will all see benefits from the improved water quality
–   *Moors for the Future Partnership's restoration and conservation of peatlands in the Pennines*: Peatlands have a vital part to play in tackling climate change, storing more carbon than all other types of vegetation in the world combined, and damage to peatlands is a significant source of carbon emissions. The Moors for the Future Partnership is already working to restore and conserve peatland in the area but needs to attract greater investment to carry out this work on a larger scale to protect more of this vital habitat. It is hoped this project will be successful in developing a range of returns, including financial, for investors[45]

## Notes

**1** OECD, "Financing Water: Investing in Sustainable Growth," https://www.oecd.org/water/Policy-Paper-Financing-Water-Investing-in-Sustainable-Growth.pdf.
**2** UN-Water, "The United Nations World Water Development Report 2018: Nature-Based Solutions for Water," (2018), https://unesdoc.unesco.org/ark:/48223/pf0000261424.
**3** Our Future Water and Climate Markets and Investment Association, "Investing in a Water-Secure Future," (2020), https://www.ourfuturewater.com/investing-in-a-water-secure-future/.
**4** OECD, "Financing Water: Investing in Sustainable Growth".
**5** Céline Kauffmann, "Financing Water Quality Management," *International Journal of Water Resources Development* 27, no. 1 (2011).
**6** R.C. Brears, "Financing Water Security," Mark and Focus, https://medium.com/mark-and-focus/financing-water-security-a7cf7caf8881.
**7** Stanford Water in the West, "Water Finance: The Imperative for Water Security and Economic Growth" (2018), https://waterinthewest.stanford.edu/sites/default/files/Water_Finance_Water_Security_Economic_Growth.pdf.
**8** High Level Panel on Water, "Making Every Drop Count: An Agenda for Water Action," (2018), https://sustainabledevelopment.un.org/content/documents/17825HLPW_Outcome.pdf.
**9** Ecologic Institute, "Economic Instruments for Water Management: Experiences from Europe and Implications for Latin America and the Caribbean," (2003), https://www.ecologic.eu/1118.
**10** Our Future Water and Climate Markets and Investment Association, "Investing in a Water-Secure Future".
**11** Ecologic Institute, "Economic Instruments for Water Management: Experiences from Europe and Implications for Latin America and the Caribbean".
**12** R.C. Brears, *Urban Water Security* (Chichester, UK; Hoboken, NJ: John Wiley & Sons, 2016).
**13** City of San Diego, "Water Billing Rates," https://www.sandiego.gov/public-utilities/customer-service/water-and-sewer-rates/water#:~:text=The%20monthly%20charges%20for%20a%20typical%20single-family%20domestic,18%20HCF%20is%20billed%20at%20%2412.488%20per%20HCF.
**14** US EPA, "Stormwater," (2015), https://www.epa.gov/sites/production/files/2015-10/documents/epa-green-infrastructure-factsheet-4-061212-pj.pdf.
**15** U.S. EPA, "Managing Wet Weather with Green Infrastructure Municipal Handbook: Incentive Mechanisms" (2009), https://www.epa.gov/sites/production/files/2015-10/documents/gi_munichandbook_incentives_0.pdf.
**16** F. A. Tasca, L. B. Assunção, and A. R. Finotti, "International Experiences in Stormwater Fee," *Water Science and Technology* 2017, no. 1 (2018).
**17** City of Guelph, "Stormwater Service Credits for Business," https://guelph.ca/living/environment/water/rebates/stormwater-service-fee-credit-program/.
**18** Stormwater Currency, "Establishing a Stormwater Volume Credit Trading Program: A Practical Guide for Stormwater Practitioners," (2019), https://www.wef.org/globalassets/assets-wef/3—resources/topics/o-z/stormwater/stormwater-institute/ar_stormwatervolumecredittrading_final_revised100919.pdf.
**19** DOEE, "Stormwater Retention Credit Eligibility and Certification Process," https://doee.dc.gov/service/stormwater-retention-credit-eligibility-and-certification-process.
**20** OECD, "Environmental Taxation: A Guide for Policy Makers," (2011), https://www.oecd.org/env/tools-evaluation/48164926.pdf.
**21** Manuel Lago et al., "Defining and Assessing Economic Policy Instruments for Sustainable Water Management," in *Use of Economic Instruments in Water Policy: Insights from International Experience*, ed. Manuel Lago, et al. (Cham: Springer International Publishing, 2015).

**22**  Helle Ørsted Nielsen et al., "Ex-Post Evaluation of the Danish Pesticide Tax: A Novel and Effective Tax Design," *Land Use Policy* 126 (2023).

**23**  Ecologic Institute, "Economic Instruments for Water Management: Experiences from Europe and Implications for Latin America and the Caribbean".

**24**  Water Supply and Sanitation Collaborative Council, "Public Funding for Sanitation – the Many Faces of Sanitation Subsidies," (2009), https://www.wsscc.org/resources-feed/public-funding-sanitation/.

**25**  Flanders Investment & Trade, "Flanders Actively Offers Ecological Investment Support," https://www.flandersinvestmentandtrade.com/invest/en/contactus.

**26**  Dionisios Latinopoulos and Eftichios S. Sartzetakis, "Using Tradable Water Permits in Irrigated Agriculture," *Environmental and Resource Economics* 60, no. 3 (2015).

**27**  Ecologic Institute, "Economic Instruments for Water Management: Experiences from Europe and Implications for Latin America and the Caribbean".

**28**  Simone Borghesi, "Water Tradable Permits: A Review of Theoretical and Case Studies," *Journal of Environmental Planning and Management* 57, no. 9 (2014).

**29**  U.S. Department of Agriculture, "The Fox Canyon Water Market: A Market-Based Tool for Groundwater Conservation Goes Live," https://www.usda.gov/media/blog/2020/05/08/fox-canyon-water-market-market-based-tool-groundwatergroundwater-conservation-goes-live.

**30**  Electric Power Research Institute, "Ohio River Basin Trading Project," https://wqt.epri.com/buy-credits.html.

**31**  Carolyn Kousky et al., "Strategically Placing Green Infrastructure: Cost-Effective Land Conservation in the Floodplain," *Environmental Science & Technology* 47, no. 8 (2013).

**32**  Carlos Eduardo Frickmann Young and Leonardo Barcellos de Bakker, "Payments for Ecosystem Services from Watershed Protection: A Methodological Assessment of the Oasis Project in Brazil," *Natureza & Conservação* 12, no. 1 (2014).

**33**  Marcela Muñoz Escobar, Robert Hollaender, and Camilo Pineda Weffer, "Institutional Durability of Payments for Watershed Ecosystem Services: Lessons from Two Case Studies from Colombia and Germany," *Ecosystem Services* 6 (2013).

**34**  OECD, "Paying for Biodiversity: Enhancing the Cost-Effectiveness of Payments for Ecosystem Services," (2010), https://www.oecd.org/env/paying-for-biodiversity-9789264090279-en.htm.

**35**  Winrock International, "Milwaukee River Pay-for-Performance Project," https://www.winrock.org/project/running-off-pollution-paying-midwestern-farmers-to-improve-water-quality/.

**36**  Delta Institute, "A New Approach to Conservation," http://deltainstitute.github.io/pay-for-performancepay-for-performance/#landscape.

**37**  World Bank, "Financing Climate Change Adaptation in Transboundary Basins: Preparing Bankable Projects," (2019), http://documents.worldbank.org/curated/en/172091548959875335/Financing-Climate-Change-Adaptation-in-Transboundary-Basins-Preparing-Bankable-Projects.

**38**  International Capital Market Association, "Green Bond Principles. Voluntary Process Guidelines for Issuing Green Bonds," (2018), https://www.icmagroup.org/green-social-and-sustainability-bonds/green-bond-principles-gbp/.

**39**  Climate Bonds Initiative, "Water Infrastructure" https://www.climatebonds.net/standard/water.

**40**  DC Water, "Green Bonds," https://www.dcwater.com/green-bonds.

**41**  OECD, "Financing Green Urban Infrastructure," in *OECD Regional Development Working Papers* (OECD, 2012).

**42**  Ibid.

**43**  Jacobs, "Jacobs Selected to Operate and Manage One of the Country's Largest Public-Private Partnerships for Wastewater Operations," https://invest.jacobs.com/investors/Press-Release-Details/2020/Jacobs-Selected-to-Operate-and-Manage-One-of-the-Countrys-Largest-Public-Private-Partnerships-for-Wastewater-Operations/default.aspx.

**44** Jason Alexandra and Curtis Riddington, *Public-Private Partnerships for Reforestation: Potential Frameworks for Investment* (Kingston, ACT: The Commonwealth of Australia, the Rural Industries Research and Development Corporation, 2007).
**45** Government of the United Kingdom, "Green Projects Given Support to Attract Private Sector Investment," https://www.gov.uk/government/news/green-projects-given-support-to-attract-private-sector-investment.

# References

Alexandra, Jason, and Curtis Riddington. *Public-Private Partnerships for Reforestation: Potential Frameworks for Investment*. Kingston, ACT: The Commonwealth of Australia, the Rural Industries Research and Development Corporation, 2007.

Borghesi, Simone. "Water Tradable Permits: A Review of Theoretical and Case Studies". *Journal of Environmental Planning and Management* 57, no. 9 (2014/09/02 2014): 1305–32.

Brears, R.C. "Financing Water Security". Mark and Focus, https://medium.com/mark-and-focus/financing-water-security-a7cf7caf8881.

____. *Urban Water Security*. Chichester, UK; Hoboken, NJ: John Wiley & Sons, 2016.

City of Guelph. "Stormwater Service Credits for Business". https://guelph.ca/living/environment/water/rebates/stormwater-service-fee-credit-program/.

Climate Bonds Initiative. "Water Infrastructure" https://www.climatebonds.net/standard/water.

DC Water. "Green Bonds". https://www.dcwater.com/green-bonds.

Delta Institute. "A New Approach to Conservation". http://deltainstitute.github.io/pay-for-performance/#landscape.

DOEE. "Stormwater Retention Credit Trading Program". https://doee.dc.gov/src.

Ecologic Institute. "Economic Instruments for Water Management: Experiences from Europe and Implications for Latin America and the Caribbean". (2003). https://www.ecologic.eu/1118.

Electric Power Research Institute. "Ohio River Basin Trading Project". https://wqt.epri.com/buy-credits.html.

Government of the United Kingdom. "Green Projects Given Support to Attract Private Sector Investment". https://www.gov.uk/government/news/green-projects-given-support-to-attract-private-sector-investment.

High Level Panel on Water. "Making Every Drop Count: An Agenda for Water Action". (2018). https://sustainabledevelopment.un.org/content/documents/17825HLPW_Outcome.pdf.

International Capital Market Association. "Green Bond Principles. Voluntary Process Guidelines for Issuing Green Bonds". (2018). https://www.icmagroup.org/green-social-and-sustainability-bonds/green-bond-principles-gbp/.

Invest in Flanders. "Flanders Actively Supports Ecological Investments". https://www.flandersinvestmentandtrade.com/invest/en/investing-in-flanders/grant-incentives/flanders-actively-supports-ecological-investments.

Jacobs. "Jacobs Selected to Operate and Manage One of the Country's Largest Public-Private Partnerships for Wastewater Operations". https://invest.jacobs.com/investors/Press-Release-Details/2020/Jacobs-Selected-to-Operate-and-Manage-One-of-the-Countrys-Largest-Public-Private-Partnerships-for-Wastewater-Operations/default.aspx.

Kauffmann, Céline. "Financing Water Quality Management". *International Journal of Water Resources Development* 27, no. 1 (2011/03/01 2011): 83–99.

Kousky, Carolyn, Sheila M. Olmstead, Margaret A. Walls, and Molly Macauley. "Strategically Placing Green Infrastructure: Cost-Effective Land Conservation in the Floodplain". *Environmental Science & Technology* 47, no. 8 (2013/04/16 2013): 3563–70.

Lago, Manuel, Jaroslav Mysiak, Carlos M. Gómez, Gonzalo Delacámara, and Alexandros Maziotis. "Defining and Assessing Economic Policy Instruments for Sustainable Water Management". In *Use of Economic Instruments in Water Policy: Insights from International Experience*, edited by Manuel Lago, Jaroslav Mysiak, Carlos M. Gómez, Gonzalo Delacámara and Alexandros Maziotis, 1–13. Cham: Springer International Publishing, 2015.

Latinopoulos, Dionisios, and Eftichios S. Sartzetakis. "Using Tradable Water Permits in Irrigated Agriculture". [In English]. *Environmental and Resource Economics* 60, no. 3 (Mar 2015 2015-02-20 2015): 349–70.

Muñoz Escobar, Marcela, Robert Hollaender, and Camilo Pineda Weffer. "Institutional Durability of Payments for Watershed Ecosystem Services: Lessons from Two Case Studies from Colombia and Germany". *Ecosystem Services* 6 (2013/12/01/ 2013): 46–53.

OECD. "Environmental Taxation: A Guide for Policy Makers". (2011). https://www.oecd.org/env/tools-evaluation/48164926.pdf.

———. "Financing Green Urban Infrastructure". In *OECD Regional Development Working Papers* OECD, 2012.

———. "Financing Water: Investing in Sustainable Growth". https://www.oecd.org/water/Policy-Paper-Financing-Water-Investing-in-Sustainable-Growth.pdf.

———  "Paying for Biodiversity: Enhancing the Cost-Effectiveness of Payments for Ecosystem Services". (2010). https://www.oecd.org/env/paying-for-biodiversity-9789264090279-en.htm.

Our Future Water and Climate Markets and Investment Association. "Investing in a Water-Secure Future". (2020). https://www.ourfuturewater.com/investing-in-a-water-secure-future/.

Skat Denmark. "Ea7.7.5 Size and Calculation of the Charge". https://skat.dk/skat.aspx?oid=1946630.

Stanford Water in the West. "Water Finance: The Imperative for Water Security and Economic Growth" (2018). https://waterinthewest.stanford.edu/sites/default/files/Water_Finance_Water_Security_Economic_Growth.pdf.

Stormwater Currency. "Establishing a Stormwater Volume Credit Trading Program: A Practical Guide for Stormwater Practitioners". (2019). https://www.wef.org/globalassets/assets-wef/3–resources/topics/o-z/stormwater/stormwater-institute/ar_stormwatervolumecredittrading_final_revised100919.pdf.

Tasca, F. A., L. B. Assunção, and A. R. Finotti. "International Experiences in Stormwater Fee". [In English]. *Water Science and Technology* 2017, no. 1 (Apr 2018 2020-03-30 2018): 287–99.

Toronto Water. "2020 Water and Wastewater Consumption Rates and Service Fees". (2019). https://www.toronto.ca/legdocs/mmis/2019/bu/bgrd/backgroundfile-139975.pdf.

U.S. Department of Agriculture. "The Fox Canyon Water Market: A Market-Based Tool for Groundwater Conservation Goes Live". https://www.usda.gov/media/blog/2020/05/08/fox-canyon-water-market-market-based-tool-groundwater-conservation-goes-live.

U.S. EPA. "Managing Wet Weather with Green Infrastructure Municipal Handbook: Incentive Mechanisms" (2009). https://www.epa.gov/sites/production/files/2015-10/documents/gi_munichandbook_incentives_0.pdf.

UN-Water. "The United Nations World Water Development Report 2018: Nature-Based Solutions for Water". (2018). https://unesdoc.unesco.org/ark:/48223/pf0000261424.

US EPA. "Stormwater". (2015). https://www.epa.gov/sites/production/files/2015-10/documents/epa-green-infrastructure-factsheet-4-061212-pj.pdf.

Water Supply and Sanitation Collaborative Council. "Public Funding for Sanitation – the Many Faces of Sanitation Subsidies". (2009). https://www.wsscc.org/resources-feed/public-funding-sanitation/.

Winrock International. "Milwaukee River Pay-for-Performance Project". https://www.winrock.org/project/running-off-pollution-paying-midwestern-farmers-to-improve-water-quality/.

World Bank. "Financing Climate Change Adaptation in Transboundary Basins: Preparing Bankable Projects". (2019). http://documents.worldbank.org/curated/en/172091548959875335/Financing-Climate-Change-Adaptation-in-Transboundary-Basins-Preparing-Bankable-Projects.

Young, Carlos Eduardo Frickmann, and Leonardo Barcellos de Bakker. "Payments for Ecosystem Services from Watershed Protection: A Methodological Assessment of the Oasis Project in Brazil". *Natureza & Conservação* 12, no. 1 (2014/06/01/ 2014): 71–78.

City of Guelph. "Stormwater Service Credits for Business." https://guelph.ca/living/environment/water/rebates/stormwater-service-fee-credit-program/.

City of San Diego. "Water Billing Rates." https://www.sandiego.gov/public-utilities/customer-service/water-and-sewer-rates/water#:~:text=The%20monthly%20charges%20for%20a%20typical%20single-family%20domestic,18%20HCF%20is%20billed%20at%20%2412.488%20per%20HCF.

DC Water. "Green Bonds." https://www.dcwater.com/green-bonds.

DOEE. "Stormwater Retention Credit Eligibility and Certification Process." https://doee.dc.gov/service/stormwater-retention-credit-eligibility-and-certification-process.

Flanders Investment & Trade. "Flanders Actively Offers Ecological Investment Support." https://www.flandersinvestmentandtrade.com/invest/en/contactus.

Nielsen, Helle Ørsted, Maria Theresia Hedegaard Konrad, Anders Branth Pedersen, and Steen Gyldenkærne. "Ex-Post Evaluation of the Danish Pesticide Tax: A Novel and Effective Tax Design." *Land Use Policy* 126 (2023/03/01/ 2023): 106549.

# Chapter 9
# Best practices and conclusion

**Abstract:** To ensure the provision of sustainable, reliable, resilient, and affordable water and water-related services that meet customers' expectations in the future, water managers will need to implement innovative water management technologies to conserve and recycle and reuse water, produce renewable energy and recover valuable nutrients from wastewater, protect and restore water quality at various scales, and improve the overall management of water resources. The financing of these technologies can be implemented through a variety of innovative financial instruments and approaches.

**Keywords:** River Basin Management, Tradable Permits, Water Quality, Best Management Practices

## Introduction

Based on the case studies, the following best practices have been identified for other regions of the world implementing innovative water management technologies that ensure the provision of sustainable, reliable, resilient, and affordable water and water-related services that meet customers' expectations in the future.

## 9.1 Conserving and recycling and reusing water

From the case studies of locations conserving and recycling and reusing water, a variety of best practices have been identified for other locations to implement:

- *Tiered Discounts and Accountability*: Cities can adopt tiered water rate systems with clear eligibility and regulation compliance criteria, require comprehensive conservation plans from participants, offer water audits, and mandate annual reporting for transparency and sustainability.
- *Acknowledgment Rates for Agriculture*: Water authorities can introduce special agricultural rates, reflecting the economic importance of farming. These rates offer reduced water costs in exchange for flexible reliability, with clear communication on usage adjustments during shortages and transparent pricing structures.
- *Adaptive Smart Metering for Conservation*: Water utilities can deploy smart meters to obtain real-time data on consumption patterns, enabling faster leak detection and more informed network understanding. Such data-driven approaches can promote water conservation, offer tailored efficiency initiatives, and facilitate

https://doi.org/10.1515/9783111028101-009

better billing options, all while aligning with broader green recovery goals and enhancing environmental stewardship.

- *Innovative Sonar Technology for Leak Detection*: Water utilities and educational institutions can collaborate to leverage cutting-edge sonar technology, utilizing fibre optics for enhanced leak detection in vast pipe networks. This approach offers agility, cost-effectiveness, and wide coverage, allowing for early-stage leak identification and rectification, resulting in significant conservation and waste reduction in urban water systems.
- *Structured Water Restrictions and Awareness*: Municipalities in water-scarce regions can enforce clear water restrictions, aligned with local climate patterns. The restrictions should emphasize the prohibition of wasteful practices. At the same time, residents should be continuously educated on simple water-saving measures. Prompt reporting and fixing of water system issues, combined with a culture of conservation and reuse, ensures sustainable resource management.
- *Product Water Efficiency Ratings*: Water utilities and authorities can initiate dynamic grading systems for water-centric appliances and fixtures. Initiating with optional labeling, the system can transition to obligatory ratings, adapting to technological progress by introducing new product categories as they emerge. It is essential for suppliers and retailers to transparently showcase these water efficiency labels, as it not only empowers consumers to opt for eco-friendly choices but also emphasizes the importance of water conservation. By periodically refining these regulations to stay in step with the latest water-saving innovations, the approach not only supports water conservation efforts but also enhances public awareness about the importance of water efficiency.
- *Comprehensive Water Education Framework*: Water utilities and educational institutions can curate platforms for in-depth exploration of water themes, from the water cycle to conservation. Tailored curricular modules, suitable for various educational levels, can be enriched with engaging games illustrating wastewater treatment or water network challenges. Embracing digital avenues, dedicated learning hubs can target different age groups, emphasizing real-world utility operations and climate impacts. Furthermore, crisis-responsive tools, like specialized educational kits, can sustain immersive learning while kindling interest in related careers.
- *Diversified Water Resilience Strategies*: Water utilities can consider integrating unconventional water sources, emphasizing increased conservation, and leveraging advanced desalination and wastewater recycling techniques. By establishing demonstration plants and educational hubs, utilities can transparently showcase technology efficacy and foster public trust. Prioritizing community-friendly applications for recycled water and thoughtful facility siting can minimize disruptions and underline commitment to sustainable water management.
- *Innovative Water Cycle Deployment*: Urban developments can optimize wastewater management by separating streams, harnessing vacuum toilets for efficient

blackwater treatment and biogas production, recycling greywater for domestic reuse, and implementing eco-centric stormwater solutions for natural replenishment and utility.

–   *Sustainable Recycled Water Utilization*: Water utilities can drive water sustainability by recycling significant volumes of wastewater annually. Adopting diversified reuse strategies, such as landscaping irrigation, residential dual reticulation for gardens and toilets, industrial consumption, and agricultural applications, ensures efficient water allocation. Tailored approaches, like dedicated irrigation schemes coupled with efficient storage solutions, can support commercial farming ventures, underscoring the potential of recycled water in bolstering both urban and agricultural water resilience.

–   *Advanced Groundwater Replenishment*: Leveraging joint ventures, utilities can revolutionize local water supply through state-of-the-art water purification systems. Harnessing multi-stage purification, from microfiltration to UV-light treatments, it guarantees potable water infusion into groundwater basins. With a progressive capacity expansion, regions can amplify their self-sustaining water frameworks, marking milestones in global water management practices.

–   *Direct Potable Reuse Excellence*: Urban centers can pioneer water sustainability by adopting innovative direct potable reuse strategies. Employing a rigorous multi-barrier treatment process, from advanced ozonation to ultrafiltration, ensures the delivery of top-tier water quality. Maintaining strict adherence to recognized water quality benchmarks and blending recycled water with natural sources enhances trust and safety. Collaborations with water treatment experts ensure long-term system reliability, while holistic public awareness campaigns foster community confidence and appreciation in modern water reuse methodologies.

## 9.2 Generating renewable energy and recovering resources from wastewater

From the case studies of locations generating renewable energy and recovering resources from wastewater, a variety of best practices have been identified for other locations to implement:

–   *Renewable Energy from Wastewater*: Utilities can capitalize on methane-rich biogas produced during wastewater treatment. By partnering with renewable energy experts, this biogas can be converted and sold in commercial pipelines, offering both environmental and financial benefits. The approach reduces carbon emissions, brings in revenue from biogas sales, and underscores a dedication to sustainable and affordable water services.

–   *Optimized Biogas-Powered Combined Heat and Power (CHP) Approach*: Wastewater facilities can harness biogas from sewage sludge to fuel CHP plants. This not only reduces costs but also boosts overall value. Advanced technologies transform this biogas into electricity and heat, significantly cutting energy consumption. Such innovations underscore a commitment to environmental sustainability and the promotion of renewable energy sources.

– *Co-Digestion for Enhanced Biogas Yield*: Co-digesting food waste with used water sludge can significantly amplify biogas production. Trials show that blending these materials can increase biogas yield up to threefold compared to treating sludge alone. This synergistic method supports sustainability goals by promoting resource recovery and energy self-sufficiency in water treatment processes.

– *Autonomous Sewage Treatment Innovation*: Employing advanced technology in sewage treatment can lead to complete energy and water self-sufficiency. Using fluidized bed incineration, waste reduction can reach up to 90%, with heat recovery generating electricity both for onsite use and surplus for public grids. Integrated desalination and rainwater recovery systems ensure independent water supply, while onsite wastewater treatment achieves zero effluent discharge. Pairing such facilities with educational and leisure amenities promotes a holistic, circular approach to waste management and community engagement.

– *Innovative Heat Pump System*: Collaborative efforts can lead to the creation of large-scale heat pumps that harness seawater and wastewater as sustainable heating sources. Powered by green electricity, such systems can significantly contribute to urban goals of transitioning to fossil-free energy. Positioned strategically, these pumps can tap into the potential of abundant natural resources for district heating. Exploring their reliability, cost-effectiveness, and adaptability can foster replication and adoption in various urban settings.

– *Floating Solar Park Innovation*: Implementing a floating solar park on a water reservoir can optimize energy production through adaptive photovoltaic modules. These modules can be adjusted based on various environmental parameters. Covering a significant portion of the water body, such installations can further sustainability goals and reduce carbon footprints. However, it is crucial to monitor the environmental impact of these installations, including potential algae growth and water quality effects. Positive outcomes can pave the way for additional similar installations on other water bodies.

– *Wind Turbine Integration in Water Treatment*: Introducing wind turbines at sewage treatment facilities can significantly reduce greenhouse gas emissions. Such initiatives lead to substantial renewable energy generation, lower carbon footprints, and energy cost savings. The initial capital investment in turbine technology can be recovered within a decade via reduced energy expenses. Collaborating with local businesses for construction can also bolster regional economic growth.

– *Hydropower Energy Recovery from Aqueducts*: Converting energy dissipation works along aqueducts into small hydropower plants optimizes unused hydraulic energy. By redirecting water through turbines, electricity is generated and then the water is returned to the aqueduct. This approach offers benefits such as renewable "green" energy production, environmental conservation, revenue from electricity sales, potential income from green certificates, job creation, and low operational costs with modern telemetry systems.

- *Enhancing Energy Efficiency in Wastewater Treatment*: Upgrading wastewater treatment facilities with modern mechanical and electrical systems improves energy efficiency. Implementing efficient aeration blowers and diffusers reduces electricity consumption. Additionally, optimizing pump configurations enhances nitrogen removal, contributing to better treated effluent quality. Advanced configurations in biological reactors further ensure improved nutrient removal and energy savings.
- *Nutrient Recovery Best Practices for Water Utilities*:
  - Water utilities can collaborate with technology providers to extract nutrients from wastewater, turning them into high-quality fertilizers. This shift offers both environmental and economic advantages. Advanced systems will prevent the accumulation of unwanted materials in pipelines, leading to significant savings on maintenance and repairs. As they embrace these innovative processes, water utilities can anticipate meeting stricter environmental regulations while ensuring sustainable and efficient operations.
  - Water utilities can integrate advanced sieving techniques to extract cellulose fibers from wastewater. This sustainable approach will transform discarded toilet paper fibers into marketable cellulose after thorough cleaning, drying, and disinfection. Such initiatives can optimize wastewater treatment processes while generating additional revenue streams from recovered materials.
  - Water utilities can harness bacteria within wastewater treatment facilities to produce PHA, a biodegradable plastic. By collaborating with research and commercial entities, utilities can refine and scale this process, transforming waste streams into valuable bioplastic resources. This approach not only offers a sustainable material alternative but also maximizes the potential of wastewater treatment.
  - Water utilities can collaborate with contractors to repurpose sewage ash into energy-efficient bricks, reducing landfill waste. By utilizing waste products from wastewater treatment, utilities can support the creation of sustainable construction materials, minimizing environmental impact and promoting circular economy practices. This approach showcases the potential to transform waste byproducts into valuable, eco-friendly resources.
  - Water utilities can innovate by recovering and reusing minerals from industrial wastewater, demonstrating the potential for zero liquid discharge. By integrating advanced technologies such as ion exchanges, dissolved air flotation, and reverse osmosis, utilities can not only reduce environmental impact but also recover valuable minerals for various industrial applications. This approach maximizes resource efficiency, promotes internal recycling of materials, and fosters a more sustainable industrial ecosystem.

## 9.3 Greening of grey water infrastructure

From the case studies of locations implementing green infrastructure solutions to manage stormwater and improve water quality, a variety of best practices have been identified for other locations to implement:

– *Rainwater Harvesting Incentive Programs*: Water utilities can implement rebate programs to promote rainwater harvesting among residents and small businesses. Offering tiered incentives based on the complexity of the harvesting systems encourages diverse uptake. By covering costs for materials, labor, and system components, utilities can motivate more users to invest in sustainable water collection practices. Additionally, mandatory educational workshops ensure participants are well-informed and can maximize the benefits of their systems.

– *Innovative Urban Stormwater Management*: Cities can turn to compact and adaptive designs, like raingarden tree pits, for efficient stormwater runoff management in dense urban areas. These systems not only clean and redirect water but also serve dual purposes such as passive irrigation for vegetation. Integrated within the streetscape, they combine environmental benefits and aesthetic appeal, all while being designed for minimal maintenance. Leveraging natural elements like substrate layers and tree roots, they offer a cost-effective solution for urban rainwater management. Regular maintenance, including cleaning, ensures the sustainability and effectiveness of these green infrastructure initiatives.

– *Green Infrastructure for Stormwater Management*: Urban areas can adopt innovative and sustainable designs, such as bioswales, to manage stormwater runoff effectively. These vegetated systems capture and treat surface water, doubling as flood control measures. Their unique designs, combining functionality and aesthetics, make them popular features in public spaces. With bio-retention systems at their base, bioswales offer both environmental benefits and community appeal. The integration of such green infrastructure promotes sustainable urban planning and provides recreational value to residents.

– *Urban Flood Mitigation and Water Quality*: Cities can implement stormwater management ponds in urban settings counteract rapid rainwater runoff. They alleviate potential flooding through temporary water storage and controlled release while improving water quality by settling sediments and contaminants. With increased impervious urban surfaces, these ponds are essential, simultaneously offering functional benefits and enhancing local biodiversity.

– *Promoting Urban Greening with Roof Incentives*: Cities can encourage sustainable urban development through green roof subsidy programs. By offering financial incentives, they can support property owners in installing green roofs, which aid in rainwater absorption, thermal insulation, and biodiversity enhancement. Key steps include assessing roof suitability with certified professionals, ensuring adherence to building rules, and providing subsidies based on the size and construction costs of the green roof. Special considerations may apply for heritage structures.

- *Adopting Sustainable Pavement Solutions*: Cities can enhance stormwater management and reduce flooding by adopting permeable interlocking concrete pavement
  (PICP) in high-traffic areas. This innovative pavement design not only promotes
  sustainability but also ensures cost-effectiveness and long-term durability, especially in climates with extreme weather conditions. The PICP installation involves
  replacing traditional asphalt with a layered system of stone and interlocking pavers, ultimately offering significant water storage capacity and increased flood
  protection for urban residents.
- *Green Infrastructure for Urban Resilience*: Cities can integrate green infrastructure
  into urban planning using technical evaluations. Combining these initiatives with
  planned construction optimizes costs. Mandating green solutions in new developments ensures sustainable growth. Furthermore, partnering with social enterprises
  for maintenance fosters community engagement and boosts local employment.
- *Multifunctional Spaces for Efficient Water Management*: Cities can adapt multifunctional spaces for stormwater management to mitigate flood risks and sewer
  overflows. Using green spaces, sports grounds, and parking areas as short-term
  retention zones during heavy precipitation events optimizes urban land use. Constructing underground storage and infiltration systems beneath facilities, such as
  stadiums, can absorb and gradually release significant volumes of rainwater,
  thereby preventing localized flooding. This integrated approach ensures both effective water management and continued community use of public spaces.

## 9.4 Protecting and restoring water quality in river basins

From the case studies of locations implementing river basin planning and other initiatives to protect and restore water quality, a variety of best practices have been identified for other locations to implement:

- *Harmonized Transnational Water Monitoring*: Water authorities can establish collaborative monitoring networks to comprehensively evaluate and track water quality,
  pollution trends, and riverine conditions in large basins. Such networks leverage national data, emphasizing harmonized methodologies, and facilitating joint monitoring
  programs. Regular assessments, shared tools, and consistent sampling frequencies
  ensure a well-structured overview of basin-wide conditions. The collected data, when
  disseminated annually or as required, aids in policy formulation and environmental
  protection strategies. Integrating systems with regional directives further ensures up-
  to-date compliance and relevant assessments.
- *Innovative Watershed-based Nutrient Trading*: Water authorities can design market-based programs that incentivize agricultural producers to implement conservation practices, offering a cost-effective alternative to conventional regulatory
  approaches. By facilitating cooperation between wastewater treatment entities
  and agriculture sectors, credit trading programs can provide sustainable revenue

sources for conservation while achieving significant nutrient reduction in rivers and streams. Preliminary economic analyses can help ascertain potential savings and benefits compared to traditional wastewater treatment upgrades.

– *Agricultural Best Management Practices (BMPs) Financing*: Water authorities can establish a dedicated loan program to provide low-interest or interest-free financing for agricultural producers aiming to implement conservation practices that reduce non-point source pollution. By offering potential principal forgiveness and flexible repayment terms based on the project's scale and lifespan, such programs can encourage widespread adoption of water and soil conservation measures. This approach also allows for the inclusion of a broad range of eligible practices, from simple tree plantings to more complex systems like manure composting and runoff management, ensuring adaptability to diverse agricultural needs.

– *Online BMP Handbook for Industrial Compliance*: Water utilities can develop an online handbook tailored for industrial sectors, offering guidance on complying with stormwater regulations. This digital platform can feature tools like monitoring guidelines, inspection forms, and customizable prevention plan templates. Regular updates, at least annually, ensure industries are always aligned with the latest regulatory requirements. The handbook can also introduce new sections as necessary, such as pollutant lists, BMP definitions, and specialized guidance materials. This approach ensures a centralized, regularly updated resource for industries, streamlining compliance processes.

– *Interactive Urban Water Management Model*: Water utilities can introduce an interactive, hands-on model using building blocks to demonstrate the importance of BMPs for managing stormwater quality and surface runoff. This engaging educational activity, adaptable for all age groups, showcases the benefits of incorporating urban water management techniques. Participants can construct a model urban environment and simulate rain events, observing the impact on runoff and pollution. By integrating features like rain gardens, water storage systems, and green infrastructure, the model visibly demonstrates reduced runoff and improved water quality, emphasizing the significance of sustainable urban planning.

– *Holistic Approach to Protecting Drinking Water Sources*: Cities can adopt a comprehensive strategy for drinking water protection by promoting sustainable agricultural and forestry practices. Engaging with local communities and stakeholders, water authorities can initiate programs that encourage the transition to organic farming and environmentally conscious forestry management. Collaborative efforts ensure that water catchment areas remain pure, reducing contaminants from farming and other human activities. By designating water protection zones around extraction sites, habitats for diverse plant and animal species are maintained, further safeguarding water quality. Such holistic approaches ensure the long-term health and sustainability of critical water sources.

## 9.5 Smart digital water management and managing customers of the future

From the case studies of locations implementing smart digital water management initiatives, a variety of best practices have been identified for other locations to implement:

- *Real-time Water Quality Monitoring of Reservoirs*: Water utilities can shift from daily manual checks to real-time monitoring of reservoirs with buoy-based systems. Using solar-powered sensors, these systems measure essential water parameters and transmit data to central platforms. Instant alerts notify operators of any deviations, ensuring swift action. This advanced approach ensures consistent water quality and proactive management.
- *Advanced Acoustic Leak Detection*: Water utilities can implement acoustic sensors to efficiently detect concealed leaks within underground water networks. These sensors can be strategically positioned along significant water mains and can identify distant non-surfacing leaks by recognizing unique sound frequencies. Utilizing a dedicated web portal, authorities can easily monitor and receive leak alerts. Optimally, sensors should operate during quieter periods, such as early mornings, to ensure accurate readings. Such technology not only helps in early leak detection but also facilitates significant water and cost savings, transitioning from reactive to proactive maintenance.
- *Innovative Pressure Management for Leak Reduction*: Water utilities can utilize advanced pressure management solutions tailored to adapt continuously to network demand variations. By doing so, pressure at critical junctures can be optimized for all network conditions. By implementing this solution, utilities can methodically modulate pressure intervals, aiming for a specified minimum pressure. Consistent and reliable operation of such systems can lead to substantial water savings, significant reductions in leakages, and overall enhanced customer service by maintaining a stable pressure regime and preventing water bursts.
- *Centralized SCADA for Infrastructure Resilience*: Water utilities can adopt centralized SCADA systems to bolster the resilience and reliability of their infrastructure. This approach enhances monitoring capabilities, offers improved cybersecurity, allows quick disaster recovery, and reduces system downtime. On-site contingencies should also be in place for immediate local control when needed.
- *Integrated GIS for Effective Emergency Response*: Water utilities can enhance emergency responses to incidents like extreme weather or pipe bursts by integrating GIS solutions. A comprehensive dashboard provides real-time incident views for tailored actions, coupled with a management tool for swift decisions and resource allocation. Post-event reviews ensure continuous improvement, while mobile apps capture real-time field data, optimizing response speed and customer service.
- *Efficient Water Usage Notifications*: Water utilities can enhance resource conservation by implementing alert systems that instantly notify users about high water usage. Through this, customers are prompted to check and fix potential leakages

promptly. Smart infrastructure and advanced metering facilitate proactive monitoring, minimizing waste, and ensuring sustainable water management.

– *Innovative Leak Detection with Smart Networks*: Water utilities can substantially enhance their infrastructure management by deploying smart network technology equipped with acoustic detection sensors. These sensors can identify and localize potential cracks or leaks in water mains. Integrating various IoT sensors, like water quality and pressure sensors, further refines the system's accuracy. Such advanced networks enable proactive detection and repair, reducing costs, minimizing community disruptions, and optimizing overall water network management.

– *Leveraging AI for Enhanced Leak Detection*: Water utilities can augment their leak detection methods by integrating AI technologies, transforming traditional acoustic loggers to smartly distinguish actual leaks from background noise. This reduces false positives and streamlines repair operations. Collaborations with tech and data science experts further refine the system's performance, ultimately leading to efficient and targeted water network maintenance.

From the case studies of locations managing customers of the future, a variety of best practices have been identified for other locations to implement:

– *Promoting Sustainable Business Practices Through Recognition*: Water utilities or cities can motivate businesses to adopt eco-friendly practices by launching recognition programs. These programs reward businesses for proactive measures like efficient water equipment usage, regular leak checks, and adherence to local environmental codes. By offering incentives such as official certifications, visibility on public platforms, and access to exclusive sustainable business networks, authorities can drive both water conservation and community engagement in environmental initiatives.

– *Engaging Communities with Conservation Contests*: Water utilities can encourage residents to embrace water-saving practices by organizing conservation-themed contests. By utilizing popular social media platforms and hashtags, such initiatives can generate community-wide engagement and raise awareness about water conservation. Rewarding the participants with attractive prizes further ensures enthusiastic participation, bringing the entire community together in the spirit of sustainability.

– *Innovative Mobile Monitoring for Water Efficiency*: Water utilities can harness the power of smartphone apps to empower customers in managing their water consumption. Such apps, with features like real-time usage tracking, leak alerts, consumption goals, and bill estimations, provide users with valuable insights and tools. This not only fosters informed decision-making but also promotes water conservation and efficient usage at an individual level.

– *Leveraging Gamification for Water Awareness*: Water utilities can capitalize on the gamification approach to foster water conservation. By integrating game elements into educational settings, they can create interactive platforms that educate and engage users about sustainable water practices. Such interactive games can address topics like preserving the environment, reducing water wastage, and preventing pollution. This strategy not only appeals to a younger audience but also challenges older users to rethink their water usage habits in an entertaining and informative way.

- *Engaging Customers through Social Media Strategies*: Water utilities can harness the power of social media platforms like Twitter to connect with and engage their customer base. By crafting diverse content, ranging from educational insights to community highlights and humorous anecdotes, they can foster a sense of community, share valuable information, and establish a relatable brand identity. Strategies can include:
  - Sharing dos and don'ts on water usage, emphasizing what can and cannot be flushed.
  - Celebrating local community events and highlights.
  - Promoting local delicacies or recipes, suggesting the use of quality water from the utility.
  - Showcasing the beauty and protection efforts of local waterways.
  - Sharing humorous and surprising finds in the water system.
  - Addressing customer concerns promptly, from outages to public interactions.
  - Advocating for sustainable practices, such as using reusable bottles.
  - Offering interactive content like polls and videos to captivate and educate audiences about daily operations and insights.
- *Effective SMS Alert Systems for Water Monitoring*: Water utilities can implement SMS alert systems, a proven method to keep residents promptly informed. Such systems can disseminate vital information about expected heavy rainfall predictions in specified regions and any coinciding tidal changes with these rain forecasts. Moreover, by monitoring water levels in essential canals and drains, especially when they breach certain thresholds, these alert systems can provide timely updates. This includes notifications about the cessation of heavy rains or when water levels recede to safer limits. Adopting such strategies ensures that the community remains safe and informed during extreme weather situations.
- *Leveraging Autonomous Drones for Reservoir Management*: Water utilities and authorities can leverage autonomous drones to optimize reservoir management. Using drones equipped with remote sensors and live-feed cameras, they can monitor water quality, track aquatic plant growth, and oversee reservoir activities. This tech-driven approach aids in identifying issues like water quality changes, plant overgrowth, or unauthorized activities. Instant alerts enable quick intervention, resulting in significant time and resource savings for reservoir management tasks.
- *Smart Water Conservation Apps*: Water utilities can introduce user-friendly mobile apps to enhance community involvement in water conservation efforts. Such apps can empower users to report water wastage by capturing real-time violations, ensuring swift action from relevant departments. Additionally, the app can educate users on current water restrictions and offer tips on water-saving practices, both

outdoors and indoors. This proactive approach not only promotes efficient water use but also fosters community collaboration towards a sustainable water future. Moreover, the app's in-built GPS can streamline issue resolution, ensuring efficient use of resources.

- *Community Involvement in Leak Detection*: Water utilities can promote active community participation in identifying and reporting water leaks to ensure rapid resolution and continued water conservation. By providing user-friendly platforms, like online maps or dedicated helplines, utilities enable easy reporting of leaks. Encouraging detailed reporting, such as precise leak locations, can further streamline the repair process. Utilities should also offer clear guidelines on leak responsibilities and support for in-home pipe bursts to maintain trust and cooperation from their customer base. This community-centric approach can help utilities maintain efficient water supply levels even as customer *bases grow.*
- *Digital Engagement for Utility Services*: Water utilities can leverage mobile apps to enhance customer interactions and service efficiency. A well-designed app should offer an intuitive account dashboard, simplified meter reading submission with integrated camera and flashlight features, secure payment options, and a dynamic mapping function for pinpointing and reporting leaks. This digital approach not only streamlines operations but also encourages community participation in water management.

## 9.6 Innovative financial instruments and approaches for water projects

From the case studies of locations implementing various innovative financial instruments and approaches to ensure the provision of sustainable, reliable, resilient, and affordable water and water-related services that meet customers' expectations in the future, a variety of best practices have been identified for other locations to implement:

- *Tiered Water Billing System Best Practices*: Water utilities or cities can introduce a tiered water billing system to promote water conservation. In such a system, customers have a base fee and are then charged incrementally based on their water consumption. As usage rises, the cost per unit of water increases. This structure not only reflects the true cost of providing water at higher consumption levels but also incentivizes users to conserve. It is important to clearly communicate the tier thresholds and rates to customers, ensuring they understand their bills and the benefits of reducing their water usage. This approach can be especially effective in regions where water scarcity is a concern.
- *Stormwater Fee Credit Initiatives*: Water utilities and authorities can bolster stormwater management by offering credits to commercial and multi-residential properties. These credits incentivize measures to reduce runoff and improve water quality. By categorizing credits like peak flow reduction and water quality

treatment, authorities address various stormwater concerns. Providing clear guidelines and a verification system ensures the initiative's effectiveness, promoting collaboration for sustainable urban development.

- *Stormwater Retention Credit Trading System*: Cities and authorities can establish stormwater retention credit (SRC) trading systems to motivate property owners to introduce green infrastructure that minimizes stormwater runoff. By generating and trading these credits, owners can both fund their projects and further the region's water quality goals. The initiative can be structured to prioritize projects offering maximum environmental benefits. Through a central agency, they can set fixed or market-based rates for the SRCs and offer resources to guide participants. Periodic application windows and clear guidelines can streamline the process, making it advantageous for both property owners and environmental conservation.
- *Pesticide Tax Based on Environmental Impact*: Water authorities can introduce a tiered pesticide tax system aimed at reducing the environmental and health impacts of pesticides. This approach taxes pesticides based on their health and environmental burden, encouraging farmers to opt for less harmful alternatives. Such a tax system takes into account various risk factors associated with each pesticide, with a higher levy placed on those with greater adverse effects. Additionally, to ensure transparency and proper tracking, farmers can be mandated to electronically submit their usage records. Revenue generated from the tax can be channelled back into the agricultural sector, offsetting other expenses, and facilitating a more sustainable approach to farming.
- *Subsidies for Eco-Friendly Investments*: Cities or authorities can provide incentive programs to encourage businesses to adopt green technologies and sustainable practices. By setting up a tiered grant system based on a predefined technology list, authorities can prioritize and fund those eco-friendly initiatives that align with regional sustainability goals. Investments in areas like energy conservation, renewable energy, and environmentally-friendly technologies can be incentivized through these grants. The grant amount can vary based on the company size and the specific technology implemented. Additionally, for forward-thinking projects that employ unlisted but innovative environmental technologies, a strategic support system can be established.
- *Tradable Water Abstraction Rights*: Water authorities can adopt a market-driven strategy to address water scarcity through a cap-and-trade system for water rights. In this model, stakeholders are allocated fixed water quantities based on historical usage, allowing them to buy or sell unused portions in an online, anonymous marketplace. The success of such a program depends on recognizing the urgency of water scarcity, ensuring stakeholder collaboration, leveraging environmental market design expertise, and securing sufficient funding and capacity.
- *Efficient Water Pollution Rights Trading*: Water authorities can implement a water quality trading program, which allows permitted dischargers to purchase nutrient reductions from other sources, optimizing cost efficiency. Such a pro-

gram's success hinges on several factors. Firstly, it's crucial to overcome any barriers to full-scale implementation. Secondly, the design of trading mechanisms should ensure ecological effectiveness and be accepted by stakeholders. Encouraging early and voluntary participation is another key aspect. Additionally, regular evaluations should be conducted to determine the program's ability to support broader ecosystem services. Lastly, comprehensive systems and protocols must be in place to ensure compliance.

– *Pay-for-Performance Water Quality Improvement*: Water authorities can adopt a pay-for-performance model to enhance agricultural water quality. This data-driven approach incentivizes farmers based on the scientifically measured outcomes of their conservation methods. The process encourages farmers to use efficient conservation strategies, predicts pollutant reductions like phosphorous using scientific models, compensates based on these predictions, then monitors and verifies the actual water quality improvements. This system offers flexibility for farmers, ensuring wastewater facilities and other stakeholders efficiently meet water quality standards.

– *Green Financing for Utilities*: Water utilities can adopt green bonds to fund eco-friendly projects. These bonds attract funds for initiatives like water quality improvements and climate resilience. Following global standards, like the Green Bond Principles, ensures proper use of funds and attracts conscious investors. With transparent reporting and potential third-party validation, utilities not only gain financial support but also highlight their dedication to environmental sustainability.

– *Innovative Public-Private Partnerships in Wastewater Management*: Cities can leverage public-private partnerships to operate, manage, and enhance their wastewater treatment facilities. Such partnerships can amalgamate operations and maintenance under expert management, offering value-added projects and potential renewals of existing structures. Partnering with experienced firms ensures plants achieve higher performance levels, promote sustainability, and move towards becoming net-zero energy facilities, subsequently mitigating environmental impacts.

– *Leveraging Public and Private Partnerships for Environmental Projects*: Water utilities and authorities can collaborate with the private sector to bolster natural environment projects. Combining public and private funding accelerates efforts to restore habitats, manage flood risks, and combat pollution. Partnering with diverse stakeholders, such as wildlife trusts and farmers' unions, ensures holistic and efficient project implementation. These initiatives offer dual benefits, addressing environmental concerns while generating sustainable financial returns.

## Conclusion

In conclusion, as the century progresses, the water sector is facing increasing pressure from a wide range of climatic and non-climatic trends that challenge its ability to provide sustainable, reliable, resilient, and affordable water and water-related services that meets customers' expectations in the future. Traditionally, the water sector has been typically slow to evolve and incorporate new innovative solutions into existing systems in response to various challenges due to a number of barriers. Nonetheless, a failure to implement innovations in water management will expose the water sector to a variety of risks to human health, the environment, and infrastructure as well as reductions in the level of service customers have come to expect. To ensure the provision of sustainable, reliable, resilient, and affordable water and water-related services that meet customers' expectations in the future, water managers will need to implement innovative water management technologies to conserve and recycle and reuse water, produce renewable energy and recover valuable nutrients from wastewater, protect and restore water quality at various scales, and improve the overall management of water resources. The financing of these technologies can be implemented through a variety of innovative financial instruments and approaches.

# Index

https://doi.org/10.1515/9783111028101-010

www.ingramcontent.com/pod-product-compliance
Lightning Source LLC
Chambersburg PA
CBHW081531220326
41598CB00036B/6399